Development of Regional Skews for Selected Flood Durations for the Central Valley Region, California, Based on Data Through Water Year 2008

By Jonathan R. Lamontagne, Jery R. Stedinger, Charles Berenbrock, Andrea G. Veilleux, Justin C. Ferris, and Donna L. Knifong

Prepared in cooperation with U.S. Army Corps of Engineers

Scientific Investigations Report 2012–5130

U.S. Department of the Interior
U.S. Geological Survey

U.S. Department of the Interior
KEN SALAZAR, Secretary

U.S. Geological Survey
Marcia K. McNutt, Director

U.S. Geological Survey, Reston, Virginia: 2012

For more information on the USGS—the Federal source for science about the Earth, its natural and living resources, natural hazards, and the environment, visit http://www.usgs.gov or call 1–888–ASK–USGS.

For an overview of USGS information products, including maps, imagery, and publications,
visit http://www.usgs.gov/pubprod

To order this and other USGS information products, visit http://store.usgs.gov

Suggested citation:
Lamontagne, J.R., Stedinger, J.R., Berenbrock, Charles, Veilleux, A.G., Ferris, J.C., and Knifong, D.L., 2012,
Development of regional skews for selected flood durations for the Central Valley Region, California, based on data
through water year 2008: U.S. Geological Survey Scientific Investigations Report 2012–5130, 60 p.

Contents

Abstract 1

Introduction 1

    Purpose and Scope 3

    Study Area Description 3

Rainfall Flood Data 5

    Basin and Climatic Characteristics 5

Cross-Correlation Model of Concurrent Flood Durations 12

Flood-Frequency Analysis 15

    Flood Frequency Based on LP3 Distribution 15

    Expected Moments Algorithm (EMA) 15

Regional Duration-Skew Analysis 24

    Standard GLS Analysis 24

    WLS/GLS Analysis 25

    Skew-Duration Analysis 25

    Use of Regional-Skew Models 33

Summary 34

References 34

Appendix 1. Unregulated Annual Maximum Rain Flood Flows for Selected Durations for
all 50 Sites in the Central Valley Region Study Area, California. 37

Appendix 2. Ancillary Tables for Regional-Skew Study in the Central Valley Region of
California 38

Appendix 3. Methodology for Regional-Skew Analysis for Rainfall Floods of Differing
Durations 48

Figures

1. Map showing study basins and site numbers for regional-skew analysis of *n*-day rainfall flood flows, Central Valley, California 4
2. Graph showing basin drainage area and site names and numbers sorted by ascending drainage area, Central Valley region, California 9
3. Graph showing mean basin elevations and site names and numbers sorted by ascending elevation, Central Valley region, California 10
4. Scatter plot showing relation between drainage area and mean basin elevation for sites draining different areas, Central Valley region, California 11
5. Scatter plot showing Fisher transformation (Z) of cross correlation between concurrent annual 1-day maximum flows and the distance between basin centroids, Central Valley region, California 13
6. Graph showing cross correlation for selected annual maximum *n*-day flows and annual peak flow in the Central Valley, California 14
7. Graph showing flood-frequency curve for maximum 3-day duration flows for the Feather River at Oroville Dam (site 13), California 16
8. Graph showing flood-frequency curve for maximum 3-day duration flows for the Trinity River at Coffee Creek (site 54), California, with no censoring and with censoring 17
9. Graph showing flood-frequency curve for maximum 3-day duration flows for the American River at Fair Oaks (site 17), California, after additional censoring 19
10. Graph showing flood-frequency curve for maximum 3-day duration flows for Putah Creek at Monticello Dam (site 44), California, after additional censoring ... 20
11. Skew coefficients of rainfall floods for all durations and for all sites in order of ascending 7-day skew for the study sites in the Central Valley region, California 21
12. Skew coefficients for 1-, 3-, 7-, 15-, and 30-day rainfall floods ordered by drainage areas in the Central Valley region, California 22
13. Skew coefficients for 1-, 3-, 7-, 15-, and 30-day rainfall floods ordered by mean basin elevation in the Central Valley region, California 23
14. At-site skews as a function of mean basin elevation with the regional model for durations of 1-day 3-days, 7-days, 15-days, and 30-days 29
15. Graph showing models of nonlinear skews for all durations in the Central Valley region, California 32

Tables

1. Site number, site name, location, record-length information, drainage area, and mean basin elevation for study basins, Central Valley region, California 6
2. Basin characteristics considered as explanatory variables and their source, Central Valley region, California 8
3. Model coefficients (a, b, and c) in equation 2b of cross correlation of concurrent annual maximum flows for selected durations 13
4. Number of study sites for differing numbers of censored low flows for each duration (see appendix 2-2 for greater details) 15
5. Mathematical models used in the regional-skew analyses 26
6. Summary of statistical results of five regional-skew models for five durations ... 27
7. Pseudo ANOVA table for the final non-linear regional-skew model for all n-day flood durations, Central Valley region, California 28
8. Variance of Prediction (VP) and Effective Record Length (ERL) for five durations as a function of mean basin elevation 33

Conversion Factors, Datums, and Acronyms and Abbreviations

Inch/Pound to SI

Multiply	By	To obtain
Length		
inch (in.)	2.54	centimeter (cm)
inch (in.)	25.4	millimeter (mm)
foot (ft)	0.3048	meter (m)
mile (mi)	1.609	kilometer (km)
Area		
acre	4,047	square meter (m^2)
acre	0.4047	hectare (ha)
acre	0.4047	square hectometer (hm^2)
acre	0.004047	square kilometer (km^2)
square foot (ft^2)	929.0	square centimeter (cm^2)
square foot (ft^2)	0.09290	square meter (m^2)
square mile (mi^2)	259.0	hectare (ha)
square mile (mi^2)	2.590	square kilometer (km^2)
Volume		
cubic foot (ft^3)	28.32	cubic decimeter (dm^3)
cubic foot (ft^3)	0.02832	cubic meter (m^3)
cubic mile (mi^3)	4.168	cubic kilometer (km^3)
Flow rate		
foot per second (ft/s)	0.3048	meter per second (m/s)
cubic foot per second (ft^3/s)*	0.02832	cubic meter per second (m^3/s)

Vertical coordinate information is referenced to North American Vertical Datum of 1988 (NAVD 88).

Elevation, as used in this report, refers to distance above the vertical datum.

Conversion Factors, Datums, and Acronyms and Abbreviations

SI to Inch/Pound

Multiply	By	To obtain
Length		
centimeter (cm)	0.3937	inch (in.)
millimeter (mm)	0.03937	inch (in.)
meter (m)	3.281	foot (ft)
kilometer (km)	0.6214	mile (mi)
Area		
square meter (m^2)	0.0002471	acre
hectare (ha)	2.471	acre
square hectometer (hm^2)	2.471	acre
square kilometer (km^2)	247.1	acre
square centimeter (cm^2)	0.001076	square foot (ft^2)
square meter (m^2)	10.76	square foot (ft^2)
hectare (ha)	0.003861	square mile (mi^2)
square kilometer (km^2)	0.3861	square mile (mi^2)
Volume		
cubic decimeter (dm^3)	0.03531	cubic foot (ft^3)
cubic meter (m^3)	35.31	cubic foot (ft^3)
cubic kilometer (km^3)	0.2399	cubic mile (mi^3)
Flow rate		
meter per second (m/s)	3.281	foot per second (ft/s)
cubic meter per second (m^3/s)	35.31	cubic foot per second (ft^3/s)

Conversion Factors, Datums, and Acronyms and Abbreviations

Acronyms and Abbreviations

USACE	U.S. Army Corps of Engineers
ANOVA	analysis of variance
AVP	average variance of prediction
ASEV	average sampling error variance
B-GLS	Bayesian generalized least squares
CAWSC	California Water Science Center
EMA	expected moments algorithm
ERL	effective record length
EVR	error variance ratio
$E[VP_{new}]$	average value of the variance of prediction at a new site
GB	Grubbs-Beck
GIS	geographic information system
GLS	generalized least squares
IACWD	Interagency Advisory Committee on Water Data
LP3	log-Pearson Type III
MBV	misrepresentation of the beta variance
MOVE.1	maintenance of variance extension, type 1 method
MSE	mean square error
NHDPlus	national hydrologic dataset
NLCD	national land-cover dataset
NL Elevation	nonlinear regional skew model
ordinary least squares	ordinary least squares
NWIS	USGS National Water Information System
OLS	ordinary least squares
PRISM	Parameter-Elevation Regressions on Independent Slopes Model
USACE	U.S. Army Corps of Engineers
USGS	U.S. Geological Survey
WLS	weighted least squares

Development of Regional Skews for Selected Flood Durations for the Central Valley Region, California, Based on Data Through Water Year 2008

By Jonathan R. Lamontagne[1], Jery R. Stedinger[1], Charles Berenbrock[2], Andrea G. Veilleux[3], Justin C. Ferris[2], and Donna L. Knifong[2]

Abstract

Flood-frequency information is important in the Central Valley region of California because of the high risk of catastrophic flooding. Most traditional flood-frequency studies focus on peak flows, but for the assessment of the adequacy of reservoirs, levees, other flood control structures, sustained flood flow (flood duration) frequency data are needed. This study focuses on rainfall or rain-on-snow floods, rather than the annual maximum, because rain events produce the largest floods in the region. A key to estimating flood-duration frequency is determining the regional skew for such data. Of the 50 sites used in this study to determine regional skew, 28 sites were considered to have little to no significant regulated flows, and for the 22 sites considered significantly regulated, unregulated daily flow data were synthesized by using reservoir storage changes and diversion records. The unregulated, annual maximum rainfall flood flows for selected durations (1-day, 3-day, 7-day, 15-day, and 30-day) for all 50 sites were furnished by the U.S. Army Corps of Engineers. Station skew was determined by using the expected moments algorithm program for fitting the Pearson Type 3 flood-frequency distribution to the logarithms of annual flood-duration data.

Bayesian generalized least squares regression procedures used in earlier studies were modified to address problems caused by large cross correlations among concurrent rainfall floods in California and to address the extensive censoring of low outliers at some sites, by using the new expected moments algorithm for fitting the LP3 distribution to rainfall flood-duration data. To properly account for these problems and to develop suitable regional-skew regression models and regression diagnostics, a combination of ordinary least squares, weighted least squares, and Bayesian generalized least squares regressions were adopted. This new methodology determined that a nonlinear model relating regional skew to mean basin elevation was the best model for each flood duration. The regional-skew values ranged from –0.74 for a flood duration of 1-day and a mean basin elevation less than 2,500 feet to values near 0 for a flood duration of 7-days and a mean basin elevation greater than 4,500 feet. This relation between skew and elevation reflects the interaction of snow and rain, which increases with increased elevation. The regional skews are more accurate, and the mean squared errors are less than in the Interagency Advisory Committee on Water Data's National skew map of Bulletin 17B.

Introduction

Flood-frequency estimates are required by engineers, land-use planners, resource managers, dam operators, and others for effective and safe use of all resources in and near California streams. Commonly, flood-frequency analyses are based on annual peak flows because peak flows on unregulated streams produce maximum flood levels. However, the flood frequency of a volume of flood flow over a duration of time—also known as the annual maximum n-day flood flow, where n represents the number of days, or duration, of flooding—is critical for the design, construction, and operation of dams and levees. Most rivers in the Central Valley region are dammed, and many of these dams are massive, such as the Oroville Dam, which has a height of 770 feet (ft), contains a volume of 77,619,000 cubic yards (yd³) of material, and provides a reservoir capacity of 3,538,000 acre-feet (acre-ft), and the

[1]Cornell University, School of Civil & Environmental Engineering, 220 Hollister Hall, Ithaca, New York 14853

[2]U.S. Geological Survey, California Water Science Center, Placer Hall, 6000 J Street, Sacramento, California 95819

[3]U.S. Geological Survey, Office of Surface Water, 12201 Sunrise Valley Drive, Reston, Virginia

Shasta Dam, which has a height of 602 ft, contains a volume of 6,270,000 yd³ of material, and provides a reservoir capacity of 4,552,000 acre-ft. Reliable estimates of *n*-day flood frequency are crucial for reservoir operation at these dam sites and also at key levee locations, where prolonged flooding could weaken structures and threaten safety. For the Central Valley region of California, the Bureau of Reclamation and the U.S. Army Corps of Engineers (USACE) determined that the frequency associated with the annual maximum 3-day rainfall flood is often the most critical flood frequency for reservoir operation. (Cudworth, 1989; U.S. Army Corps of Engineers, 1997 and 2002; National Reasearch Council, 1999; and Hickey and others, 2002). The maximum 3-day rainfall flood is critical for dam release rates and associated flood-control storage space in many reservoirs. The frequency associated with the annual maximum 7-day rainfall flood is also important because it represents back-to-back 3-day duration rainfall floods, which are not uncommon in the Central Valley region of California.

Consequently, the regional-skew analysis for this study focuses on annual maximum *n*-day flood-duration flows. Specifically, regional skew was determined for the annual maximum 1-day, 3-day, 7-day, 15-day, and 30-day flood-duration flows. Results from this study complement a regional-skew analysis for annual peak flow in California recently completed by Parrett and others (2011).

This study was initiated through a collaborative effort between the USACE, and the U.S. Geological Survey (USGS) California Water Science Center (CAWSC). Currently, USACE and the California Department of Water Resources are reassessing flood hazards in the Central Valley region of California. The Sacramento and San Joaquin River basins lie within this region, where river and tributary flooding historically have threatened several large population centers, including the City of Sacramento. Many of the levees that compose the extensive flood protection system in the Sacramento and San Joaquin River basins are being upgraded, or have been targeted for rehabilitation or upgrading. To ensure that levee enhancements are designed using the best available flood-frequency estimates, a new regional-skew analysis was conducted. This study is an extension of a previous flood-duration study by USACE (Hickey and others, 2002; U.S. Army Corps of Engineers, 2002). The previous study used only at-site rainfall flood data because rainfall generally produces the largest floods in the Central Valley region. Accordingly, this study also used only rainfall *n*-day flood data and did not include snowmelt *n*-day flood data.

Bulletin 17B from the Hydrology Subcommittee of the Interagency Advisory Committee on Water Data (1982), hereinafter referred to as Bulletin 17B, recommends the use of log-Pearson Type III distribution when estimating flood frequency at gaged sites. The shape of the log-Pearson Type III distribution depends on the standard deviation and skew coefficient. The precision of flood-frequency estimates depends largely on the precision of the estimated skew coefficient, particularly for extreme floods, which are of greatest interest (Griffis and others, 2004). The skew coefficient is difficult to estimate from small sample sizes because it is very sensitive to the presence of outliers or unusual observations. For this reason, Bulletin 17B recommends the use of a weighted skew coefficient that is a weighted average of a combination of the skew coefficient estimated from the flood data at a site and the regional-skew coefficient. The weighted skew coefficient is used to estimate the flood quantiles of interest, and the weights assigned to the at-site and regional skew depend on the relative precision of the two skew estimators.

Since the publication of Bulletin 17B in 1982, there have been significant advances in statistical methodologies and computing technology that supports regional hydrologic regression assessments. Studies by Reis and others (2005), Weaver and others (2009), Feaster and others (2009), and Gotvald and others (2009) have shown that Bayesian Generalized Least Squares (GLS) regression provides an effective statistical framework for estimating regional-skew coefficients for annual peak flows as well as their precision. Bayesian GLS regression provides more precise regional-skew coefficients for annual peak flows than the National skew map provided in Bulletin 17B. Bayesian GLS regression was adapted for use in the California regional-skew analysis for annual peak flows reported by Parrett and others (2011). The study reported here uses a similar methodology to that of Parrett and others (2011) with some modifications. The regional-skew analysis for peak flows in California and this study are both based on a hybrid weighted least squares and generalized least squares (WLS/GLS) procedure, which was needed because of the large cross-correlations among concurrent flood flows at stream sites in California.

An important first step in the regional-skew analysis is the estimation of the skew coefficient of the logarithms of the flood data for each site included in the study. Many of the sites had flood data that contained low outliers or zero flow observations, both of which require special treatment in order to calculate skew coefficients that are characteristic of the largest observations. The expected moments algorithm (EMA), which has been shown to more efficiently account for censored observations than Bulletin 17B recommended procedures, was used to fit a log-Pearson Type III distribution to each of the flood records in this study (Cohn and others, 1997, 2001; Griffis and others, 2004). Unregulated, annual maximum rainfall flood data for 1-day, 3-day, 7-day, 15-day, and 30-day durations for each study site were provided by USACE, and basin characteristics for each study site were provided by the USGS.

This study was a collaboration between Cornell University and the USGS. Previous collaborations produced a new regional skew for annual peak flows for the southeastern region of the United States, including parts of Virginia, Tennessee, Alabama, North Carolina, South Carolina, and Florida (Weaver and others, 2009; Feaster and others, 2009; and Gotvald and others, 2009), and for much of California (Parrett and others, 2011).

Purpose and Scope

The primary purposes of this report are to (1) present the results of regional-skew analysis for rainfall floods for selected *n*-day durations for the Central Valley and adjacent regions of California, and (2) to describe the newly developed hydrologic regression methodology that was used. Fifty sites of interest (streamgages and major dams) to USACE in the region were used in the study. A database of unregulated, annual maximum rainfall floods for durations of 1-day, 3-days, 7-days, 15-days, and 30-days at these sites was provided by USACE and is presented in appendix 1. The *n*-day flood-duration flow is the maximum avarge discharge of any consecutive *n*-day period in a water year for a site. Because dam sites were included in the regional-skew analysis, unregulated, flood-duration data at those sites had to be synthesized. For most of these sites, the daily unregulated discharge was synthesized from reservoir storage or diversion records. For one dam site, however, the maintenance of variance extension, type 1 method (MOVE.1; Helsel and Hirsch, 1992), was used to synthesize flow data. A database of basin characteristics was also developed for the basin upstream from each site. These characteristics are presented in appendix 2.

The new EMA methodology was used to compute moments of the logarithms of discharge for the LP3 distribution to determine a station skew at each site to be used in the regional-skew analysis. A visual censoring procedure for low outliers and zero flows was utilized. The number of censored observations and zero flows for each site is given in appendix 2. A newly developed Bayesian hybrid WLS/GLS regression procedure was used to develop the regional-skew model for each duration. The approach is described in appendix 3. Finally, diagnostic statistics commonly reported for Bayesian GLS, including values of leverage and influence for each site, are presented in appendix 3.

Study Area Description

The stream sites used in this regional-skew study of the Central Valley region of California are shown in figure 1. Originally, 55 sites were considered for this study, but only 50 sites were employed in the final analysis. Three of the five sites were dropped (site 2, Clear Creek near Igo; site 21, Lost Banos Creek at Los Banos Dam; and site 27, Littlejohn Creek at Farmington Dam) because reliable flood records were unavailable. Two sites (site 22, Orestimba Creek near Newman, and site 29, Cosgrove Creek near Valley Springs) were dropped because their basin hydrology is uncharacteristic of the Central Valley study region and particularly uncharacteristic of the major dam sites of interest to USACE.

Roughly two-thirds of the sites included in this study drain the western slopes of the Sierra Nevada Range, located along California's eastern border. Streams draining this region account for the majority of the flow into the Sacramento and San Joaquin Rivers. Peak elevations generally increases in the Sierra Nevada with decreasing latitude. Basins in this region with a mean elevation greater than about 4,000 ft experience significant annual snowpack, which probably affects annual flood characteristics. Also, this region experiences rain-on-snow events, where warm temperatures cause precipitation to fall as rain, which causes the snowpack to melt and runoff rapidly (Parrett and others, 2011; Mount, 1995). Flood data resulting from these rain-on-snow events are considered to be rainfall floods for this report.

The remaining one-third of the study sites drain the Coastal Ranges, which parallels California's Pacific coast. Peak elevations in the Coastal Ranges generally are much lower than in the Sierra Nevada, and basins in the Coastal Ranges generally do not accumulate significant snowpack compared to basins in the Sierra Nevada. Annual maximum floods in the Coastal Ranges are generally caused by large winter rainstorms (Parrett and others, 2011). Hydrologic conditions in basins in this region vary widely from north to south, but generally the northern Coastal basins have more annual rainfall than the southern Coastal basins.

Parrett and others (2011) discuss the influence of the complicated interaction of rain and snow in forming annual maximum floods. They noted that annual peak floods in basins that have a mean elevation lower than 4,000 ft are usually caused by rain and that the influence of rain and snow interactions is greater with increasing elevation. Annual peak floods in basins with mean elevations greater than 8,000 ft are most often caused by snowmelt runoff. Data from floods caused by snowmelt runoff were not used in this study.

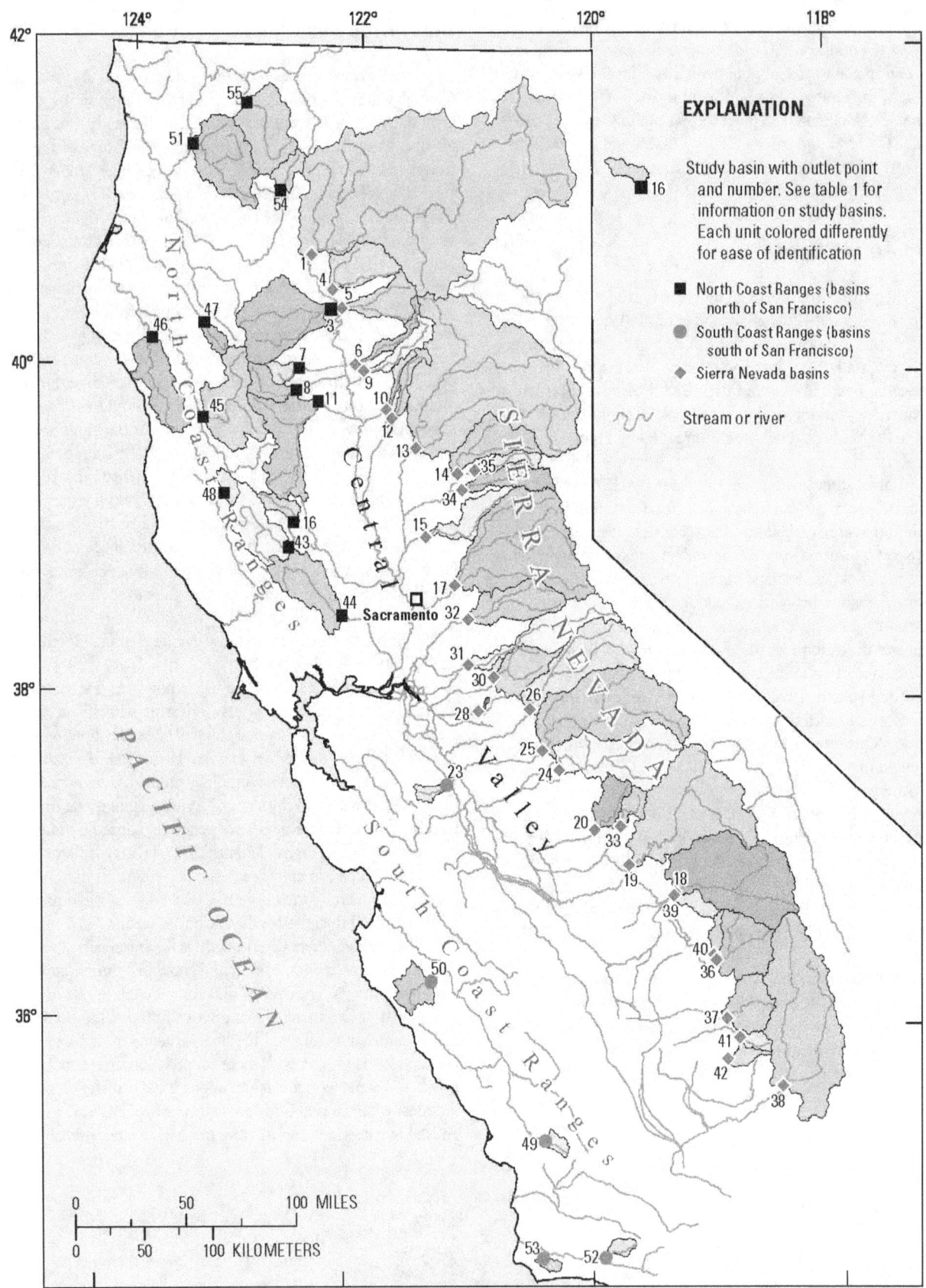

Figure 1.   Study basins and site numbers for regional-skew analysis of *n*-day rainfall flood flows, Central Valley, California. (See table 1 for site names corresponding to the site numbers shown.)

Rainfall Flood Data

Unregulated, annual maximum flow data resulting from rainfall for the 1-day, 3-day, 7-day, 15-day, and 30-day durations were provided by USACE for each of the 50 sites used in the study (appendix 1). These sites have record lengths ranging from 30 to 113 years, and all but four sites have records through water years 2008 or 2009. Site information for each of the study basins is listed in table 1. Of the 50 sites, 28 experienced no significant regulation during the period of record. For the remaining 22 sites, daily unregulated-flow data were synthesized from daily regulated-flow records and reservoir storage or diversion records. From the synthesized, daily unregulated-flow data for each year, USACE determined the annual maximum n-day floods from rainfall for each year. The USACE methodology for generating an annual series of n-day floods from rainfall is as follows: (1) obtain daily mean flow data for a site; (2) if necessary, augment the daily mean flow data using reservoir storage or diversion data to obtain synthesized daily unregulated-flow data; (3) if necessary, for each year of daily unregulated-flow data, remove daily data predominantly due to snowmelt runoff; and (4) for each year of resulting daily unregulated flows from rainfall, calculate the annual maximum value of daily flow averaged over each n-day duration.

Twenty-eight of the fifty sites in this study also were included in the earlier Sacramento–San Joaquin Comprehensive Study (hereafter referred to as "Comp Study"), which published annual maximum values of n-day flood flows from rainfall through water years 1998 or 1999 (U.S. Army Corps of Engineers, 2002). These records were extended through water years 2008 or 2009 except for four sites (appendix 1). Seven sites were extended through water year 2009. The Calaveras River at New Hogan Dam (site 30) record was extended from 1908 to 1964 by applying the MOVE.1 technique (Maintenance Of Variance-Extension, type 1; Helsel and Hirsch, 1992) to streamflow records at one downstream and three upstream gages and the change in storage records from old Hogan Dam.

Runoff events in the Sierra Nevada can be characterized by two overlapping statistical populations: rainfall events and snowmelt events. For basins with mean elevations lower than about 3,000 ft, runoff is essentially all from rainfall. As basin elevations increase above about 3,000 ft, the effects of snowpack and snowmelt on runoff increase. Although the snowpack that melts during a rainfall flood event could have accumulated over several months, the snowmelt runoff was still considered part of the rainfall flood in this study.

In basins with mean elevations above about 8,000 ft, significant snowpack can remain late into the spring and early summer. In these watersheds, annual maximum flows can be the result of rainfall runoff, snowmelt runoff, or a combination of these events. Twenty sites were identified by visual inspection of the unregulated, daily flow series to have both rainfall and snowmelt flood flows in the record of annual maximum n-day flows. As described above,

daily flows resulting from snowmelt were removed from the annual records. Bulletin 17B (Interagency Advisory Committee on Water Data, 1982, p. 16) provides guidance for event separation: "Separation by calendar periods in lieu of separation by events is not considered hydrologically reasonable unless the events in the separate periods are clearly caused by different hydrometeorologic conditions." Previously, in the Comp Study, rainfall and snowmelt flood populations were separated by visually inspecting the unregulated-flow hydrograph for each water year. The inspection was augmented by snowpack and temperature data. Analyst judgment was used to determine the beginning of the snowmelt season for each year. In most water years, this served as the date of segregation. If the annual maximum flow was the result of a late season rainfall event that occurred after the start of snowmelt, the date of segregation was adjusted to include the late season event in the rainfall population. The separation procedure used by USACE in this study was consistent with the Comp Study procedure.

The Kern River at Isabella Dam (site 38) and the Kaweah River at Terminus Dam (site 36; fig. 1) have basins where snowmelt runoff represents an annual base flow or minimum flow due to snowmelt that can be subtracted from the rain flood record. Base flow was estimated graphically as a lower bound for the frequency curves. Base flows of 150 cubic feet per second (ft^3/sec) for the Kern site and 60 ft^3/sec for the Kaweah site were subtracted from rainfall flood series for all five durations before log-Pearson Type III (LP3) distributions were fit to the adjusted datasets. This corresponds to a flow-separation procedure that allowed the statistical analysis to focus on the magnitude of the larger annual maximum rainfall flood series for each duration.

Basin and Climatic Characteristics

The suite of basin characteristics for each of the 50 sites in the regional duration-discharge skew analyses was derived from various national geographic information system (GIS) databases, including the National Hydrologic Dataset (NHDPlus), National Land-Cover Dataset (NLCD), and the Parameter-Elevation Regressions on Independent Slopes Model (PRISM) climatic dataset for data from 1970 to 2000. Table 2 describes the explanatory GIS variables and their data sources. The same quality-assurance standards used to create the GIS database of basin characteristics in the report by Parrett and others (2011) were used in this study. Differences between the older, manually measured drainage areas in the NWIS (National Water Information System) database and the drainage areas determined from the GIS database were identified. Differences in drainage area for the two databases were never more than 10 percent and were within the precision of both databases. Thus, the accuracy of the basin characteristics derived from the digital GIS database was judged to be sufficient for this study.

Table 1. Site number, site name, location, record-length information, drainage area, and mean basin elevation for study basins, Central Valley region, California.

[**Abbreviations**: NAVD 88, North Americal Vertical Datum of 1988; NC, located in the North Coast Ranges north of San Francisco; S, located in the Sierra Nevada; SC, located in the South Coast Ranges south of San Francisco]

Site number	Site name	Location of site	Period of record	Number of years of record	Drainage area (square miles)	Mean elevation (feet above NAVD 88)
1	Sacramento River at Shasta Dam	S	[1]1932–2008	[1]77	6,403	4,571
3	Cottonwood Creek near Cottonwood	NC	1941–2008	68	922	2,221
4	Cow Creek near Millville	S	1950–2008	59	423	2,251
5	Battle Creek below Coleman Fish Hatchery	S	1941–2008	68	361	4,074
6	Mill Creek near Los Molinos	S	1929–2008	80	131	3,962
7	Elder Creek near Paskenta	NC	1949–2008	60	93	2,998
8	Thomes Creek at Paskenta	NC	1921–1996	76	204	4,146
9	Deer Creek near Vina	S	1912–1915, 1921–2008	92	209	4,199
10	Big Chico Creek near Chico	S	1932–2008	77	72	3,111
11	Stony Creek at Black Butte Dam	NC	[1]1901–2008	[1]108	740	2,416
12	Butte Creek near Chico	S	1931–2008	78	148	3,717
13	Feather River at Oroville Dam	S	1902–2008	107	3,591	5,031
14	North Yuba River at Bullards Bar Dam	S	1941–2008	68	489	4,899
15	Bear River near Wheatland	S	1906–2008	103	292	2,250
16	North Fork Cache Creek at Indian Valley Dam	NC	[1]1931–2008	[1]78	120	2,627
17	American River at Fair Oaks	S	1905–2008	104	1,887	4,356
18	Kings River at Pine Flat Dam	S	1896–2008	113	1,544	7,634
19	San Joaquin River at Friant Dam	S	1904–2008	105	1,639	7,046
20	Chowchilla River at Buchanan Dam	S	[1]1922–1923, 1931–2008	[1]80	235	2,152
23	Del Puerto Creek near Patterson	SC	1966–2009	44	73	1,835
24	Merced River at Exchequer Dam	S	[1]1902–2008	[1]107	1,038	5,473
25	Tuolumne River at New Don Pedro Dam	S	1897–2008	112	1,533	5,882
26	Stanislaus River at New Melones Dam	S	[1]1916–2008	[1]93	904	5,663
28	Duck Creek near Farmington	S	1980–2009	30	11	249
30	Calaveras River at New Hogan Dam	S	1908–1943, 1951–2008	96	372	1,991
31	Mokelumne River at Camanche Dam	S	[1]1905–2008	[1]104	628	4,918
32	Cosumnes River at Michigan Bar	S	1908–2008	101	535	3,064
33	Fresno River near Knowles	S	1912, 1916–1990	76	134	3,201
34	South Yuba River at Jones Bar	S	1941–1948, 1960–2008	57	311	5,362

Table 1. Site number, site name, location, record-length information, drainage area, and mean basin elevation for study basins, Central Valley region, California.—Continued

[**Abbreviations**: NAVD 88, North American Vertical Datum of 1988; NC, located in the North Coast Range north of San Francisco; S, located in the Sierra Nevada; SC, located in the South Coast Range south of San Francisco]

Site number	Site name	Location of site	Period of record	Number of years of record	Drainage area (square miles)	Mean elevation (feet above NAVD 88)
35	Middle Yuba River below Our House Dam	S	1969–1971, 1975–2008	37	145	5,365
36	Kaweah River at Terminus Dam	S	1960–2009	50	560	5,635
37	Tule River at Success Dam	S	1959–2008	50	392	3,975
38	Kern River at Isabella Dam	S	1894–1907, 1909–1915, 1917–2009	114	2,075	7,198
39	Mill Creek near Piedra	S	1958–2009	52	115	2,637
40	Dry Creek near Lemoncove	S	1960–2009	50	76	2,668
41	Deer Creek near Fountain Springs	S	1969–2009	41	83	3,989
42	White River near Ducor	S	1943–1953, 1971–2005	46	91	2,443
43	Cache Creek at Clear Lake	NC	1922–2008	87	527	2,004
44	Putah Creek at Monticello Dam	NC	1931–2008	78	567	1,327
45	Middle Fork Eel River near Dos Rios	NC	1966–2008	43	745	3,685
46	South Fork Eel River near Miranda	NC	1941–2008	68	537	1,726
47	Mad River above Ruth Reservoir near Forest Glen	NC	1981–2008	28	94	3,705
48	East Fork Russian River near Calpella	NC	1942–2008	67	92	1,630
49	Salinas River near Pozo	SC	1943–1983	41	70	2,211
50	Arroyo Seco near Soledad	SC	[1]1902–2008	[1]107	241	2,494
51	Salmon River at Somes Bar	NC	1912–1915, 1928–1929, 1931–2008	84	751	4,261
52	Santa Cruz Creek near Santa Ynez	SC	1942–2008	67	74	3,355
53	Salsipuedes Creek near Lompoc	SC	1942–2008	67	47	920
54	Trinity River above Coffee Creak near Trinity Center	NC	1958–2008	51	148	5,340
55	Scott River near Fort Jones	NC	1942–2008	67	662	4,333

[1] The period of record and number of years of record could be less than the given value for some flood durations.

Table 2. Basin characteristics considered as explanatory variables and their source, Central Valley region, California.

[**Abbreviations:** cm, centimeter; m, meter; na, not applicable; °, degrees; ', minutes; ", seconds]

Name	Description	Data source
BASINPERIM	Perimeter, in miles	30-m DEM, NHDPlus elev_cm grid http://www.horizon-systems.com/NHDPlus/
RELIEF	Relief, in feet	30-m DEM, NHDPlus elev_cm grid http://www.horizon-systems.com/NHDPlus/
ELEV	Mean basin elevation, in feet	30-m DEM, NHDPlus elev_cm grid http://www.horizon-systems.com/NHDPlus/
DRNAREA	Basin drainage area, in square miles	na
ELEVMAX	Maximum elevation, in feet	30-m DEM, NHDPlus elev_cm grid http://www.horizon-systems.com/NHDPlus/
MINBELEV	Minimum elevation, in feet	30-m DEM, NHDPlus elev_cm grid http://www.horizon-systems.com/NHDPlus/
LAKEAREA	Percent of area covered by lakes and ponds	2001 National Land Cover Database (NLCD)– land Cover http://www.mrlc.gov/nlcd2001.php
EL6000	High Elevation Index– percent of basin area with elevation greater than 6,000 feet	30-m DEM, NHDPlus elev_cm grid http://www.horizon-systems.com/NHDPlus/
OUTLETELEV	Elevation at outlet, in feet	30-m DEM, NHDPlus elev_cm grid http://www.horizon-systems.com/NHDPlus/
RELRELF	Basin relief divided by basin perimeter, in feet per mile	na
DIST2COAST	Distance in miles from basin centroid to coast along a line perpendicular to eastern California border	na
BSLDEM30M	Average basin slope, in percent	30-m DEM, NHDPlus elev_cm grid http://www.horizon-systems.com/NHDPlus/
FOREST	Percentage of basin covered by forest	2001 National Land Cover Database (NLCD)– percent Canopy http://www.mrlc.gov/nlcd2001.php
IMPNLCD01	Percentage of basin covered by impervious surface	2001 National Land Cover Database (NLCD)– percent Impervious http://www.mrlc.gov/nlcd2001.php
PRECIP	Mean annual precipitation, in inches	800M resolution PRISM 1971–2000 data http://www.prism.oregonstate.edu/products/
JANMAXTMP	Average maximum January temperature, in Fahrenheit	800M resolution PRISM 1971–2000 data http://www.prism.oregonstate.edu/products/
JANMINTMP	Average minimum January temperature, in Fahrenheit	800M resolution PRISM 1971–2000 data http://www.prism.oregonstate.edu/products/
CENTROIDX	X coordinate of the centroid, in decimal degree	na
CENTROIDY	Y coordinate of the centroid, in decimal degree	na
OUTLETX	X coordinate of the basin outlet, in meters [1]	na
OUTLETY	Y coordinate of the basin outlet, in meters [1]	na
NL Elev	Nonlinear function of elevation	Computed from the mean basin elevation

[1] Project parameters: 1st standard parallel = 29°30'00"; 2nd standard parallel = 45°30'00"; central meridian = –96°00'00"; base latitude = 23°00'00"; false easting = 0.000; false northing = 0.000.

Appendix 2–1 lists all of the basin characteristics for the 50 sites used in the regional duration-skew analysis. For three basins (sites 1, 13, and 20), some basin characteristics could not be determined, but drainage area, mean basin elevation, relief, maximum basin elevation, minimum basin elevation, and basin centroid were available for all sites. Figure 2 shows drainage area for each site sorted in ascending order. The Sacramento River at Shasta Dam (site 1) has the largest drainage area at 6,403 square miles (mi²), and Duck Creek near Farmington (site 28) has the smallest drainage area at 11 mi². Most of the study basins range in size from 100 to 1,000 mi².

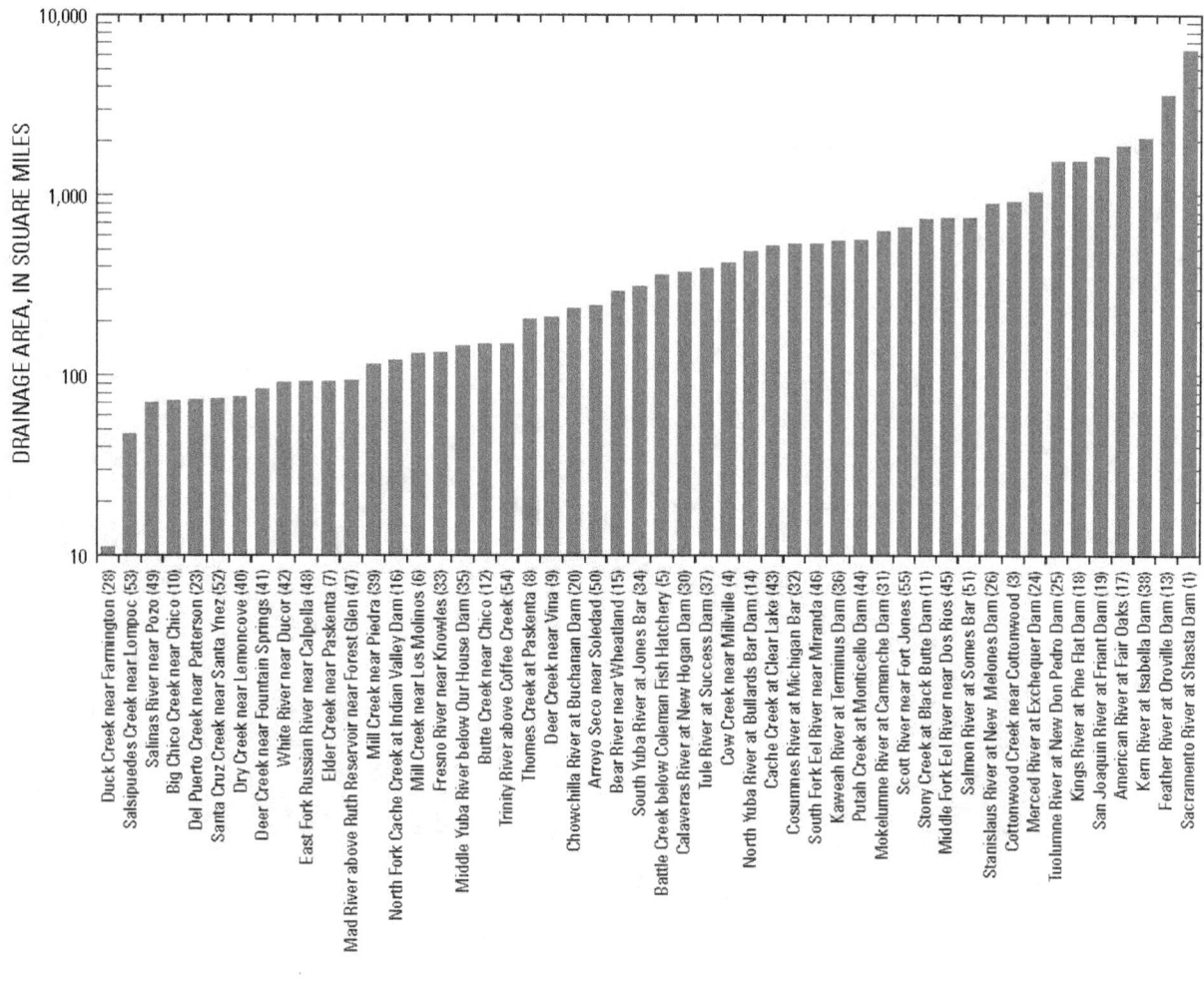

Figure 2.   Basin drainage area and site names and numbers sorted by ascending drainage area, Central Valley region, California.

Mean basin elevation for each site, sorted in ascending order, is shown in figure 3. Mean basin elevation ranged from 249 to over 7,600 ft. The Kings River at Pine Flats Dam (site 18; mean basin elevation of 7,634 ft), the San Joaquin at Friant Dam (site 19; mean basin elevation of 7,046 ft) and the Kern River at Isabella Dam (site 38; mean basin elevation of 7,198 ft) have the highest mean basin elevations in the study area. These three basins drain westerly from around Mt. Whitney (elevation of 14,505 ft and tallest mountain in the continuous US) in the southern portion of the Central Valley (fig. 1). At 249 ft, Duck Creek near Farmington (site 28) has the lowest mean basin elevation in the study area, as well as having the smallest drainage area. Whereas most basins drain either the Sierra Nevada or Coastal Range mountains, Duck Creek near Farmington drains very low-lying lands on the valley floor of the Central Valley. The mean basin elevation is a simple one-dimensional measure of the elevation of these basins. Flood hydrology in the basins depends upon a complex interaction of the drainage area, elevation, precipitation, basin orientation and rain shadow effects of the mountains, and soil conditions.

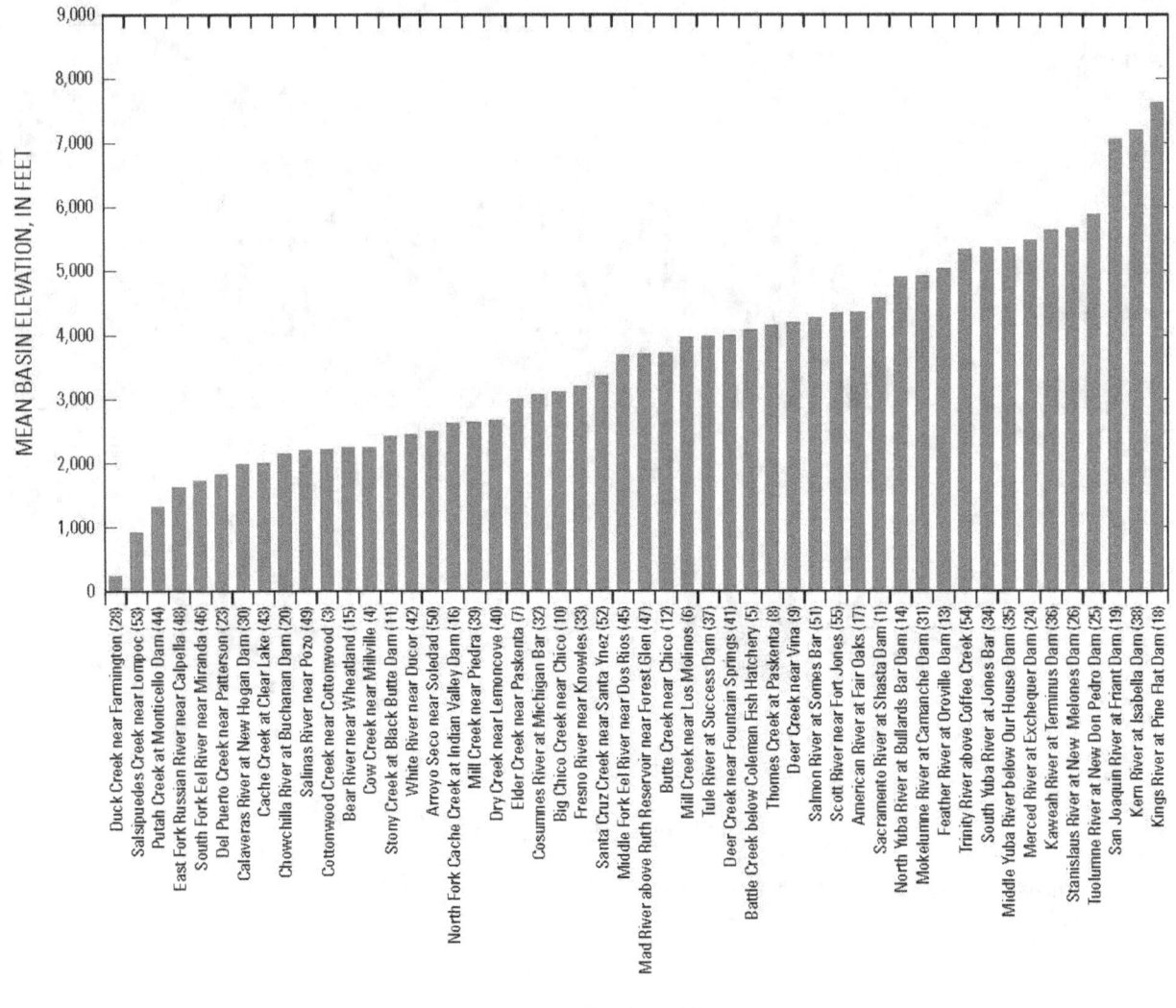

Figure 3.   Mean basin elevations and site names and numbers sorted by ascending elevation, Central Valley region, California.

The drainage area and mean basin elevation for sites in different areas are shown in figure 4. The highest and largest basins are located in the Sierra. The South Coast basins tend to be smaller and slightly lower than basins in the North Coast and the Sierra regions. Figure 4 indicates that Duck Creek near Farmington (site 28), which is a very small and low basin on the western Central Valley floor, is an obvious outlier.

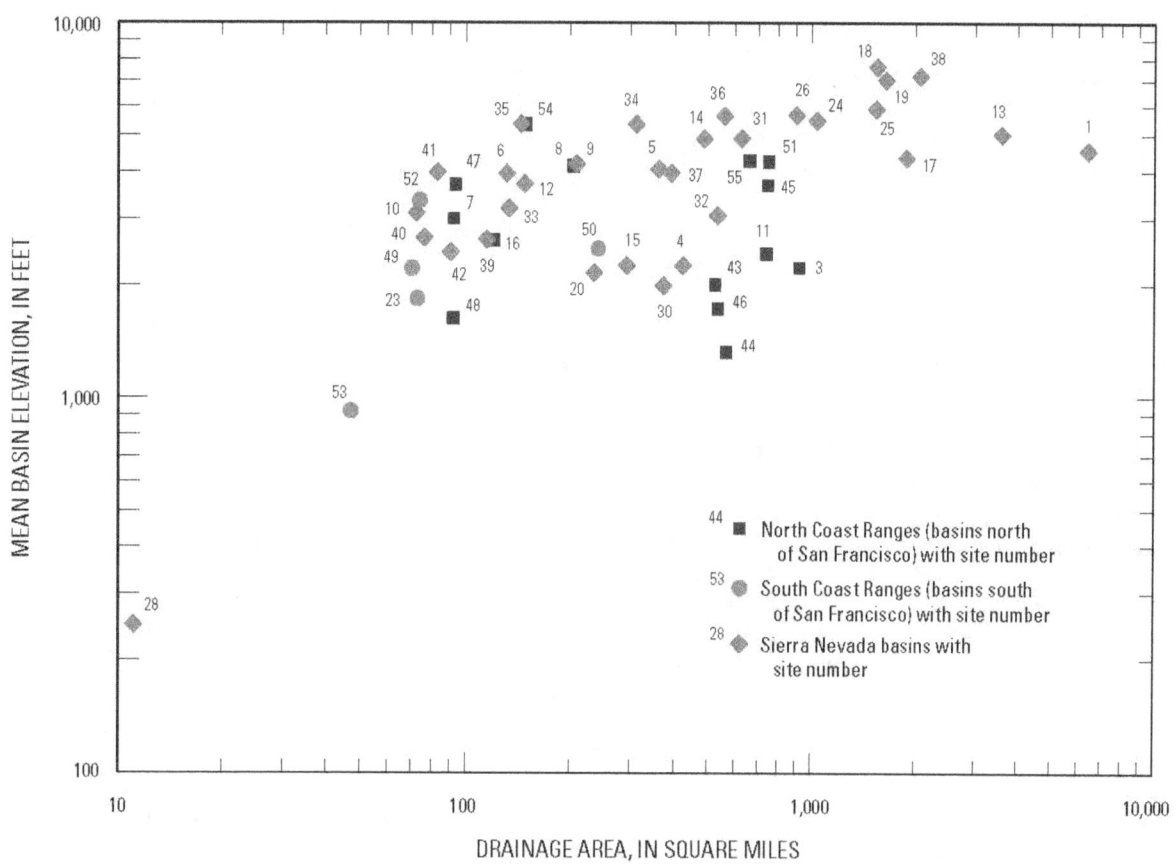

Figure 4.   Relation between drainage area and mean basin elevation for sites draining different areas, Central Valley region, California.

Cross-Correlation Model of Concurrent Flood Durations

An important step in regional-skew studies is the development of an appropriate model for the cross correlation of annual maximum n-day flood-duration flows at different sites. These cross-correlation models are used to estimate the cross correlation among skew coefficients at the different sites. Cross correlation is important, particularly when assessing model uncertainty, because sites with highly correlated concurrent annual maxima do not represent independent samples.

Basins that are spatially close to one another probably experience similar hydrologic conditions, which increases cross correlation among the concurrent n-day flood flows. For example, the three study basins with the greatest average elevations (site 18, Kings River at Pine Flat Dam; site 38, Kern River at Isabella Dam; and site 19, San Joaquin River at Friant Dam) drain the western slopes in and around Mt. Whitney in the southern Sierra Nevada Range and usually experience the same regional storms. Similarly, basins that are farther apart probably experience relatively different hydrologic conditions, resulting in lower cross correlation among the annual floods for each duration. Thus, the cross correlation between flood flows in two basins can be estimated as a function of the distance between basin centroids. Previous studies have tried functional relations with other explanatory variables, such as the ratio of drainage areas of two basins, but have generally found that functions of distance are the most useful (Gruber and Stedinger, 2008; Parrett and others, 2011).

Cross correlations between longer duration floods are expected to be greater than for shorter duration floods. The shorter duration floods are more likely to be linked to spatially limited variations in storm intensities, whereas longer duration floods are most likely linked to longer duration, spatially extensive storm systems with less variable average intensities. While 1-day and 30-day floods at a site are often linked to the same storm system, averaging runoff over the longer duration tends to dampen the effects of spatial and temporal variability on the 30-day flood.

A cross-correlation model for each n-day duration flood was developed by using sites that had at least 50 years of concurrent records with every other site. A logit model using the Fisher Z Transform,

$$Z = 0.5 \log\left[(1+r)\big/(1-r)\right] \qquad (1)$$

provided a convenient transformation of the $[-1, +1]$ range of the sample correlations r_{ij} to the $(-\infty, +\infty)$ range. The adopted model for the cross correlations of concurrent annual maximum n-day discharges at two sites, which used the distance (d_{ij}) between centroids of basins i and j as the only explanatory variable, is as follows:

$$\rho_{ij} = \frac{\exp(2Z_{ij})-1}{\exp(2Z_{ij})+1} \qquad (2a)$$

where

$$Z_{ij} = a + \exp\left(b - c \times d_{ij}\right) \qquad (2b)$$

This model is similar to those used in the earlier California and Southeast United States annual maximum flood studies. Ordinary least squares regression was used to fit the cross-correlation model for each duration. Table 3 presents the parameters for the 1-day, 3-day, 7-day, 15-day, and 30-day flood-duration models. Figure 5 shows the fitted Fisher Z transformed cross-correlation model and the distance between basin centroids for 628 station pairs for the 1-day flood-duration flows.

Figure 6 displays the fitted correlation functions for each of the five durations in this study together with the cross-correlation function for annual peak flows reported by Parrett and others (2011). The cross correlations in this study of rainfall n-day duration floods were significantly greater than the cross correlations of annual peak flows (Parrett and others, 2011). Cross correlations increased with increasing duration and with decreasing distance between basin centroids.

Table 3.   Model coefficients (a, b, and c) in equation 2b of cross correlation of concurrent annual maximum flows for selected durations.

Duration (days)	Coefficients		
	a	b	c
1	0.378	0.147	0.00605
3	0.384	0.222	0.00562
7	0.377	0.261	0.00496
15	0.355	0.315	0.00458
30	0.414	0.283	0.00480

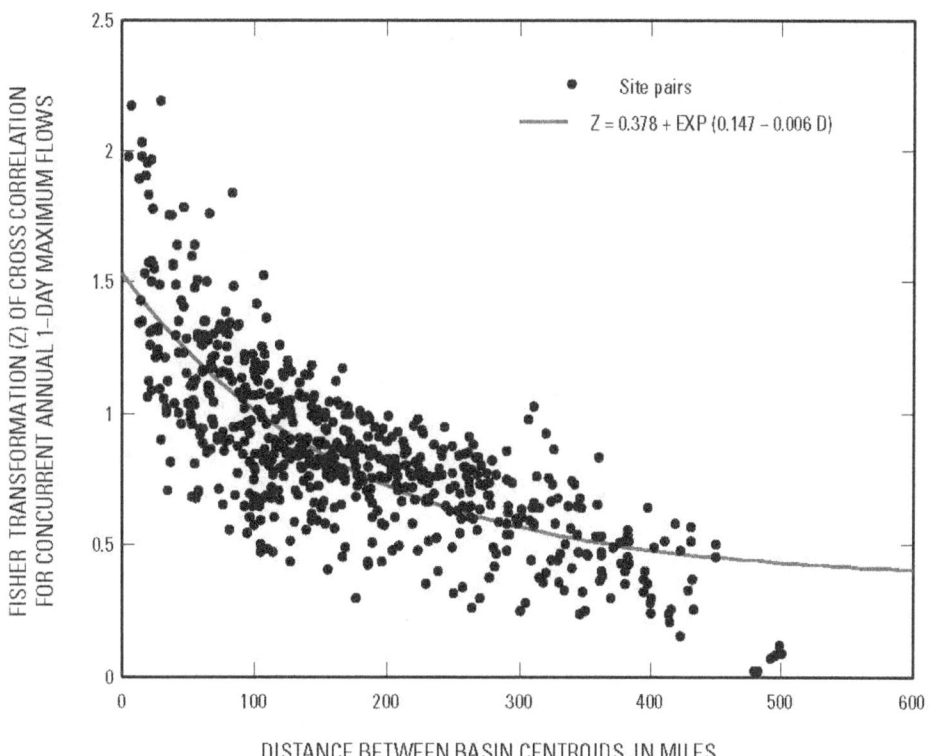

Figure 5.   Fisher transformation (Z) of cross correlation between concurrent annual 1-day maximum flows and the distance between basin centroids, Central Valley region, California.

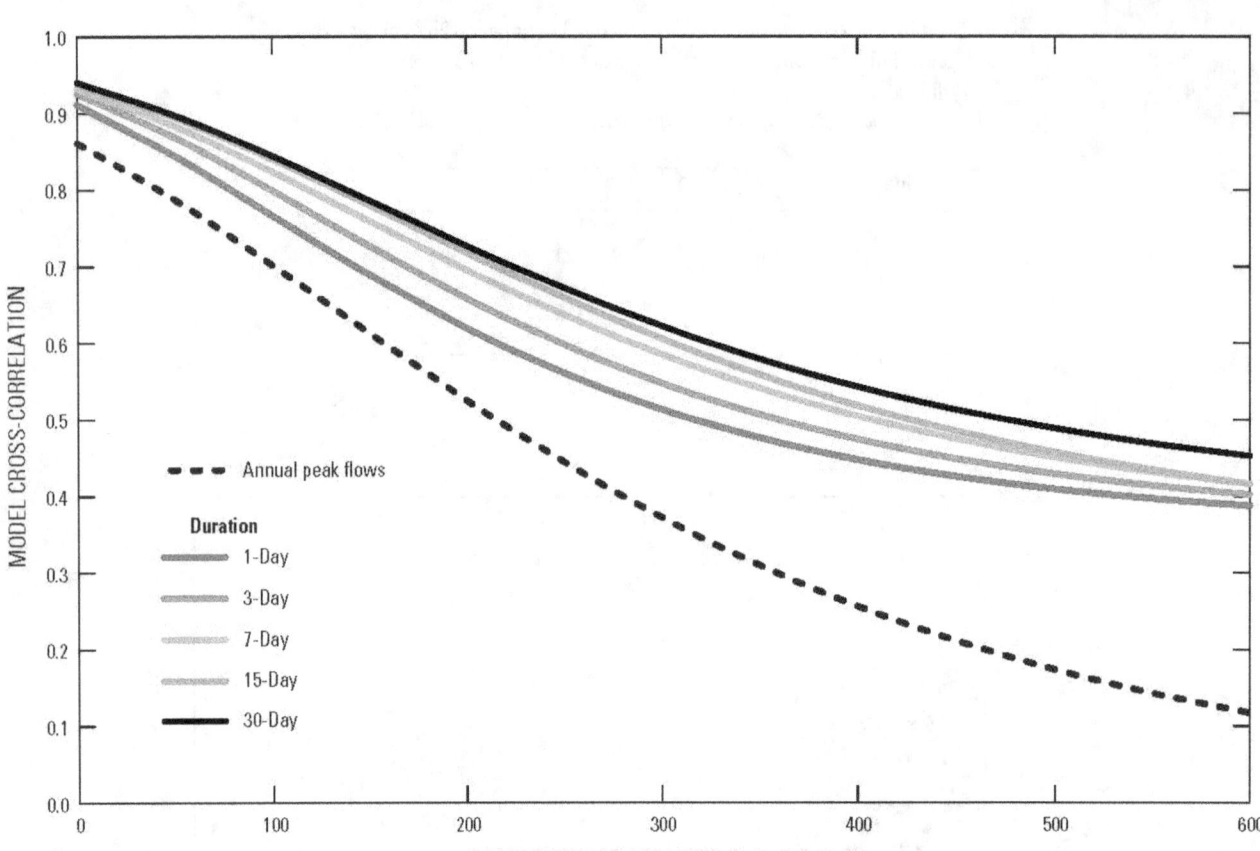

Figure 6.   Cross correlation for selected annual maximum *n*-day flows and annual peak flow in the Central Valley, California. Annual peak flows from Parrett and others (2011).

Flood-Frequency Analysis

Flood-frequency analysis for gaged sites generally involves fitting a probability distribution to the series of annual maximum discharges. Flood-frequency quantiles are often reported as T-year discharges, where T is a recurrence interval corresponding to the average number of years between annual flood discharges of the same or greater magnitude. Alternatively, flood quantiles are also reported in terms of their annual exceedance probability. Annual exceedance probability for a T-year discharge is 1/T. The annual exceedance probability is often multiplied by 100 and expressed in terms of annual percent chance of exceedance. Thus, a 100-year flood discharge has an annual exceedance probability of 0.01 and a 1.0-percent chance of exceedance in any year. Bulletin 17B (Interagency Advisory Committee on Water Data, 1982), provides guidelines and procedures for flood-frequency analysis used by federal agencies in the United States. As recommended by Bulletin 17B, the log-Pearson Type III distribution (LP3) was used for flood-frequency analyses in this study.

Flood Frequency Based on LP3 Distribution

For this study, the annual maximum n-day flows caused by rainfall were fit to the LP3 distribution. Bulletin 17B recommends fitting the Pearson Type III (P3) distribution by using the method-of-moments estimators of the mean, standard deviation, and skew coefficient of the logarithms of the flows which is the log-Pearson Type III (LP3) distribution. Given these three parameters, various flood quantiles can be computed by using the following equation:

$$\log Q_T = \overline{X} + K_T S \qquad (3)$$

where

Q_T is the flood quantile, in cubic feet per second, with recurrence interval T, in years;

$\overline{X}$ is the estimated mean of the logarithms of the annual n-day flows;

K_T is a frequency factor based on the skewness coefficient and recurrence interval, T, years; and

S is the estimated standard deviation of the logarithms of the annual n-day flows.

Rather than using the sample-skew coefficient calculated from the flow record directly, Bulletin 17B recommends the use of a weighted average of the sample skew from the flow record and a regional-skew coefficient. The weight given to each value is proportional to its relative precision, expressed as mean squared error. The precision of the sample-skew estimator is a function of the record length at a site, so that the longer the period of record, the more weight that is given to the sample skew relative to the regional value.

Expected Moments Algorithm (EMA)

This study used the expected moments algorithm (EMA) for fitting the LP3 distribution (Cohn and others, 1997; Cohn and others, 2001; England and others, 2003a,b; Griffis and others, 2004; Parrett and others, 2011). The EMA method for calculating the LP3 moment estimators is more robust and efficient than those described in Bulletin 17B when various forms of censored flows are part of the flow record, including zero flows; low outliers; "below-threshold" observations, wherein the flow is described as Q is less than Q_0 for some threshold Q_0; and historical or paleoflood flows. For this study, the only forms of censored flows in the records were zero flows and low outliers. Bulletin 17B includes a Grubbs-Beck (GB) test for determining if an observed flow should be classified as a low outlier. Most low outliers are flows that are significantly less than other flows in a flood record. A very important concern is that unusually low flows in a flood record can cause the fitted flood distribution to have a very negative skew, which, in turn, causes the distribution to diverge from the largest floods in the record. Because large floods are the main concern in flood-frequency studies, it is often advisable to censor the smaller flood flows to allow the fitted distribution to correctly describe the risk of large flood flows. Table 4 summarizes the number of sites that had differing numbers of censored low flows for each n-day duration. Table 2–2 in appendix 2 contains a detailed accounting of censoring and zero flows for each site included in the study.

Table 4. Number of study sites for differing numbers of censored low flows for each duration (see appendix 2-2 for greater details).

Number censored	1-Day	3-Day	7-Day	15-Day	30-Day
0	16	16	16	16	16
1	28	28	28	28	28
2	1	2	2	2	2
3	1	0	0	0	0
4	2	2	2	2	3
5	0	1	1	1	0
>5	2	1	1	1	1
Total	50	50	50	50	50

About one-third of the 50 sites included in this study had neither zero-flow observations nor censored low outliers. The flood-frequency curve for annual maximum 3-day flows for the Feather River at Oroville Dam (site 13), displayed in figure 7, is an example of an LP3 curve fit for a site that had no censored observations.

The other two-thirds of the study sites had at least one zero flow or other flood observation that was identified as a low outlier by using the GB test. The GB test is a 10-percent significance test for the smallest observation in a log-normally distributed sample, which corresponds to an LP3 distribution with zero skew (Interagency Advisory Committee on Water Data, 1982). The GB test identified low outliers in about half of the records in this study. In many instances, the GB criterion worked well and markedly improved the LP3 curve fit. In some cases, additional censoring was advisable; appendix 2-2 summarizes the actions taken for this study.

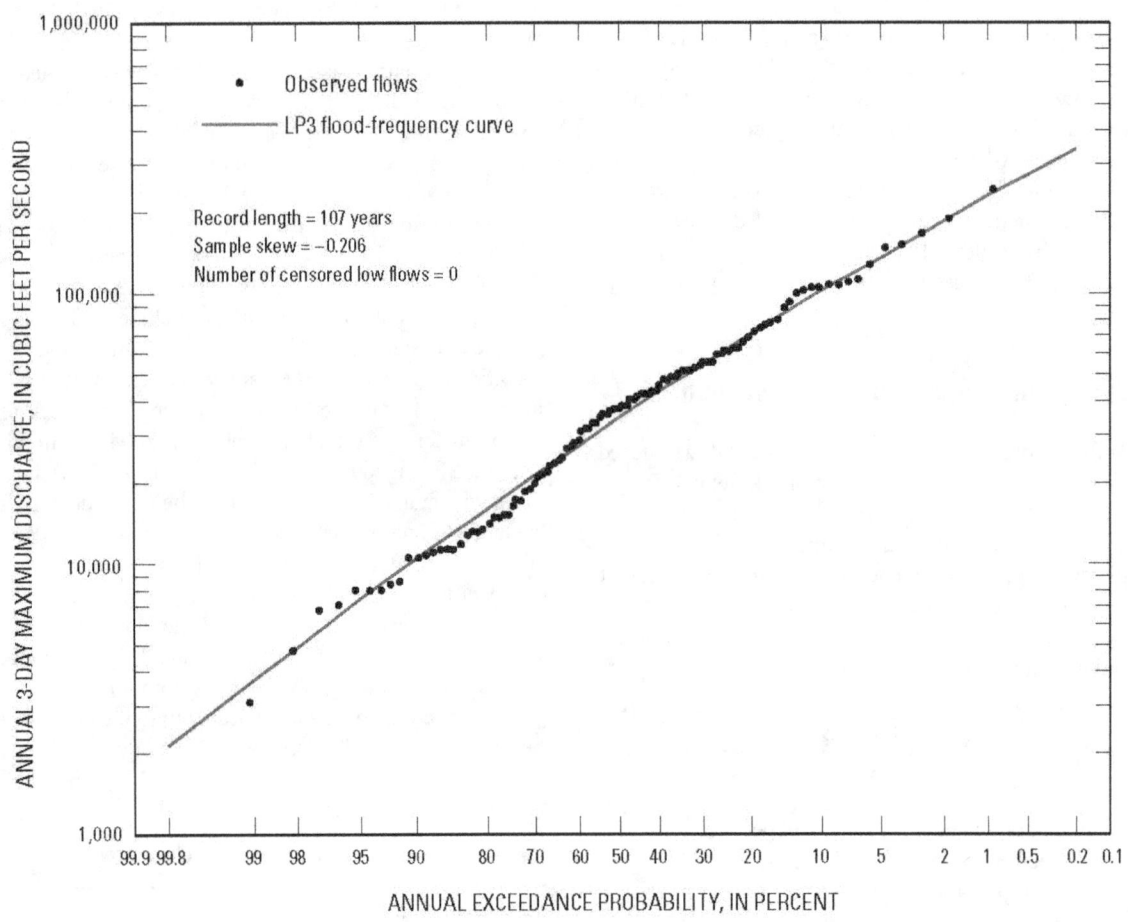

Figure 7. Flood-frequency curve for maximum 3-day duration flows for the Feather River at Oroville Dam (site 13), California.

The Trinity River above Coffee Creek near Trinity Center (site 54) provides an example of the effects caused by a single low outlier. The GB test identified one low outlier in the flow record for the annual maximum 3-day flood flows at this site. As shown in figure 8A, the calculated skew based on all the flood flows (−0.254) produces a flood-frequency curve that is concave and does not fit the three largest observations. After censoring the smallest observation, as indicated by the GB test, the calculated skew is less negative (−0.061), and the frequency curve fits the largest observations better (fig. 8B).

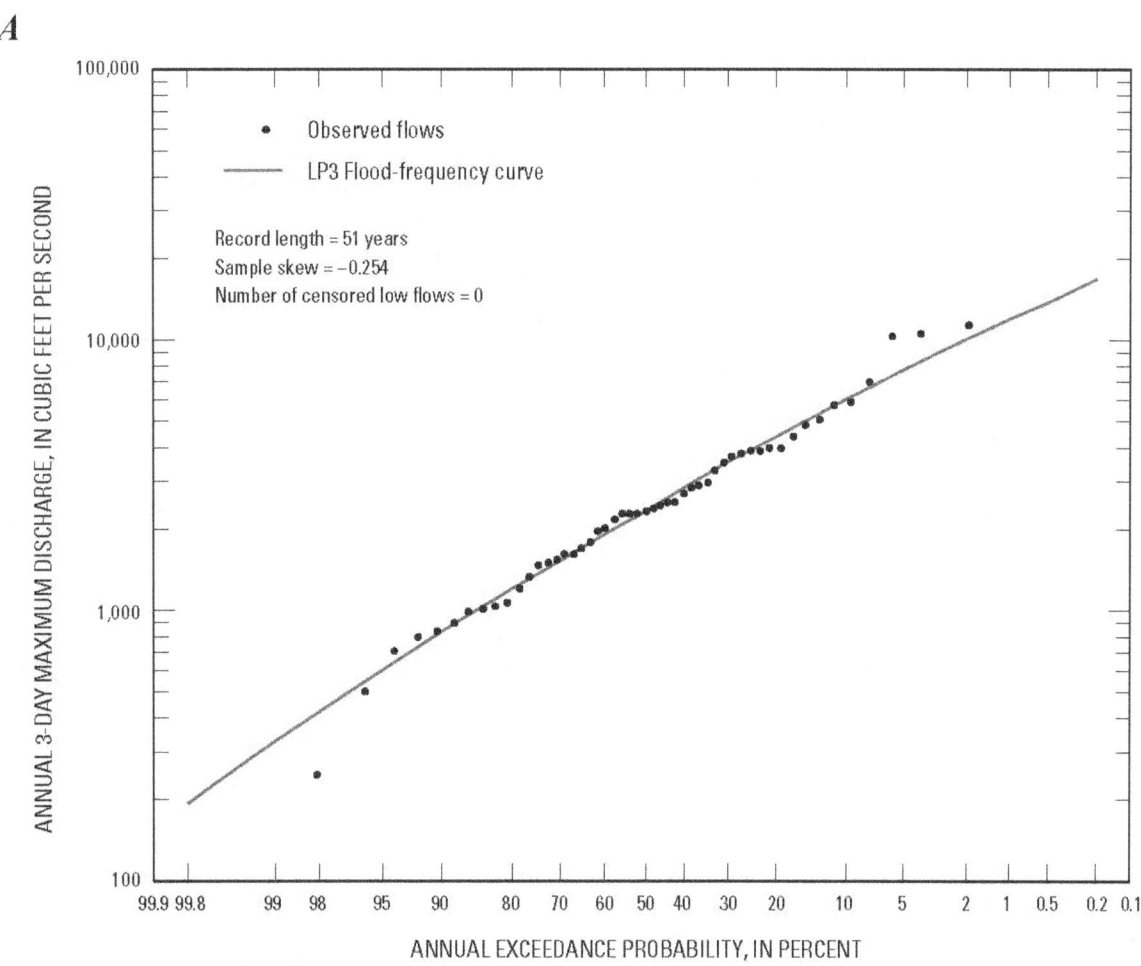

Figure 8. Flood-frequency curve for maximum 3-day duration flows for the Trinity River at Coffee Creek (site 54), California, (A) with no censoring and (B) with censoring.

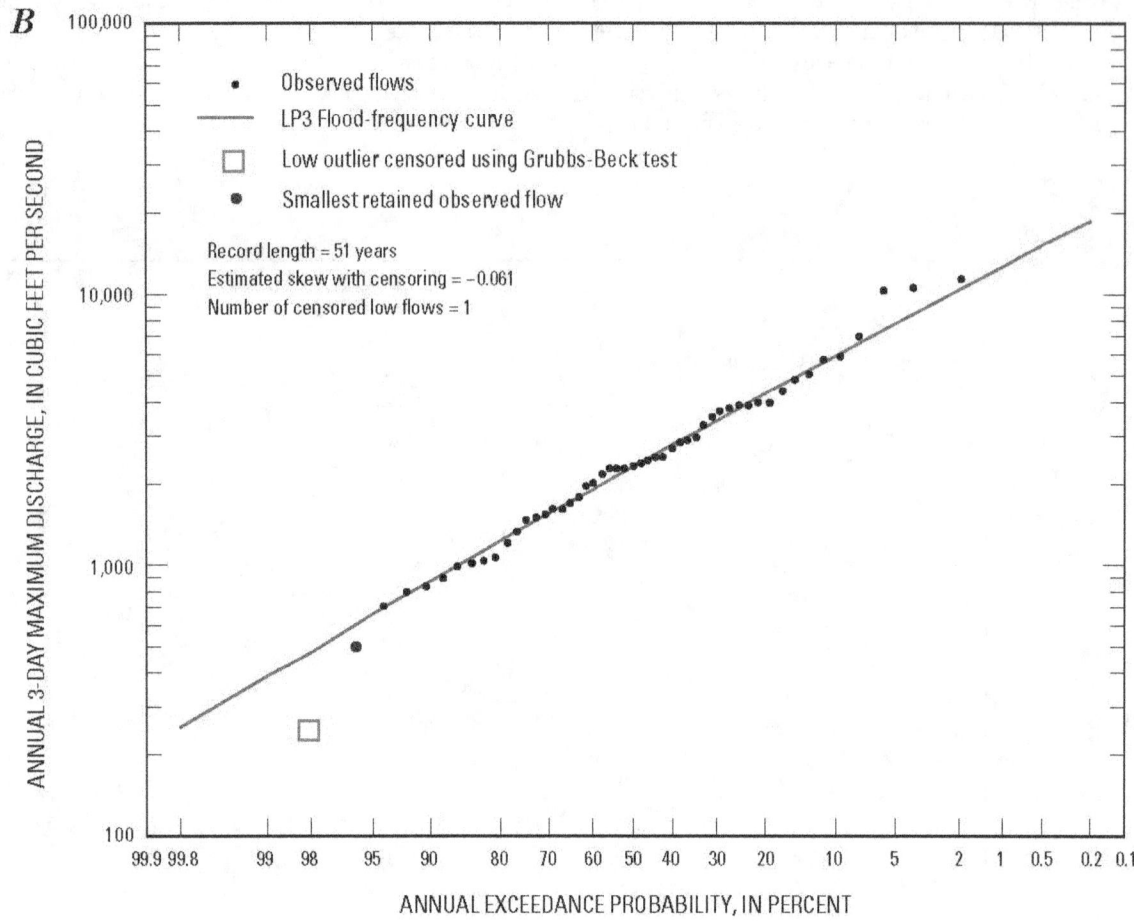

B

Legend:
- Observed flows
- LP3 Flood-frequency curve
- Low outlier censored using Grubbs-Beck test
- Smallest retained observed flow

Record length = 51 years
Estimated skew with censoring = –0.061
Number of censored low flows = 1

Y-axis: ANNUAL 3-DAY MAXIMUM DISCHARGE, IN CUBIC FEET PER SECOND

X-axis: ANNUAL EXCEEDANCE PROBABILITY, IN PERCENT

Figure 8.—Continued

Although the GB test provides a reasonable procedure for determining all low outliers at many sites, visual inspection of the plotted frequency curves is often required to identify and censor other low outliers that could significantly affect the LP3 curve fits to the greatest observed flood flows. Additional low-outlier censoring was most common for basins in drier regions. Additional low outliers were visually identified in about 25 to 30 percent of the flow records, depending on the duration. For example, the GB test failed to identify any low outliers in the annual maximum 3-day flood record for the American River at Fair Oaks (site 17), but visual inspection of the flood-frequency curve clearly revealed one low outlier (fig. 9). Censoring that observation by describing it as less than the smallest retained observation improved the fit of the frequency curve to the greatest observed flood flows.

Another concern with flood data for the American River at Fair Oaks (site 17) was consistency in censoring the flood record across durations. The apparent low outlier visually identified in the annual maximum 3-day flow record was from 1977. While the GB test failed to identify the maximum 3-day flow in 1977 as a low outlier, the GB test did identify the 1977 flood event as a low outlier for other n-day durations. Annual flood maxima for various n-day durations are usually produced by the same storm; thus, an outlier for one duration was generally treated as an outlier at other durations in this study. For the American River example, censoring the same annual flood flow for all durations maintained consistency in the fitting of the LP3 curves to flows for all durations.

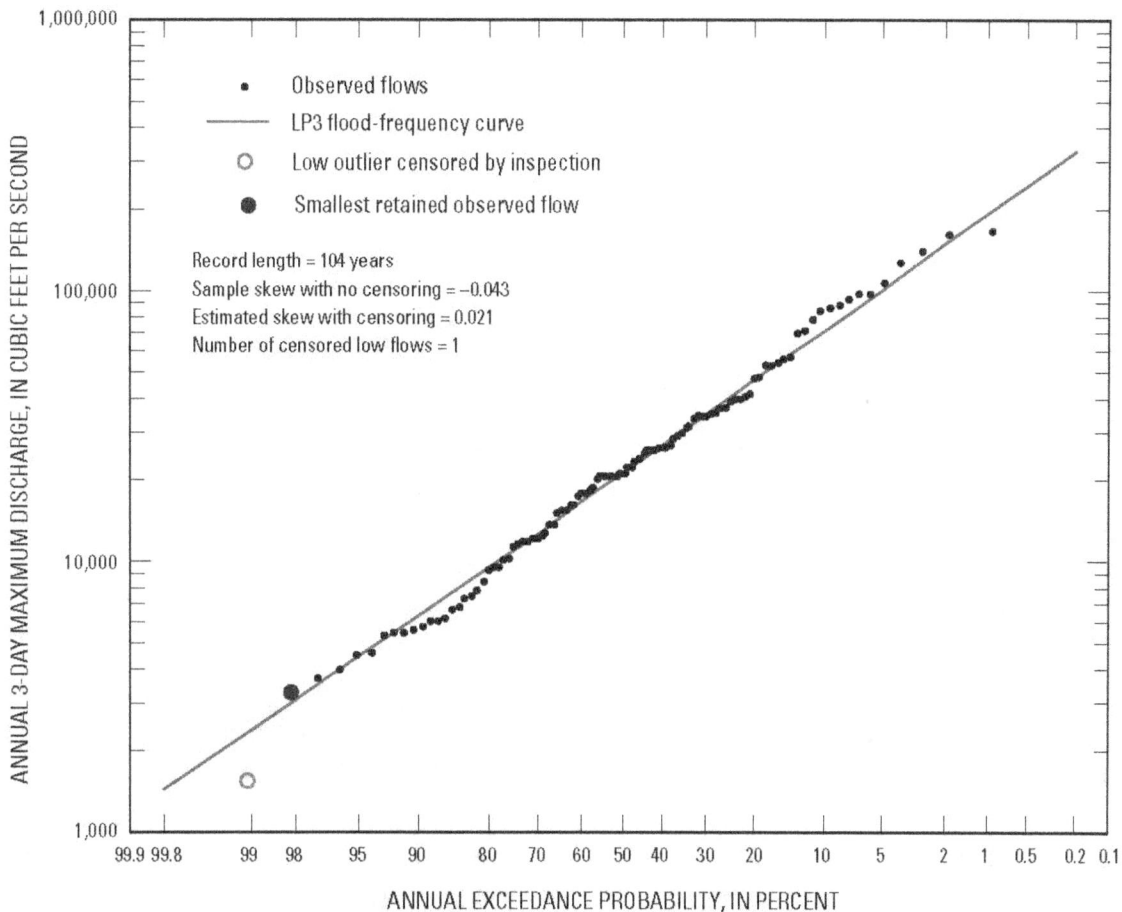

Figure 9. Flood-frequency curve for maximum 3-day duration flows for the American River at Fair Oaks (site 17), California, after additional censoring.

Additional visual censoring in this study typically involved the removal of just one or two observations that were not identified by the GB test. However, a few sites such as Putah Creek at Monticello Dam (site 44), required more extensive censoring because the LP3 frequency curve was unable to provide a good fit to both the smallest and largest observations in the record. The censoring process entailed censoring the lowest observations one-by-one until there was an adequate fit of the frequency curve to the larger flood observations. Censoring increases the mean square error (MSE) of the at-site skew-coefficient estimate, and extensive censoring—used to improve the at-site LP3 curve fit to larger observations—can critically effect the weight placed on station skew in a regional-skew analysis because the weight given to each estimated skew depends inversely upon its MSE. The most extensive censoring in this study was carried out for flow records at Putah Creek at Monticello Dam (site 44). The GB test identified only one low-outlier in the annual maximum 1-day flood flows at this site, but visual inspection indicated a significantly improved LP3 curve fit when 11 additional flows were censored (fig. 10). The EMA censoring threshold for all 12 low outliers was the smallest retained flood observation, which represents the upper bound that is assigned to all of the lower censored observations.

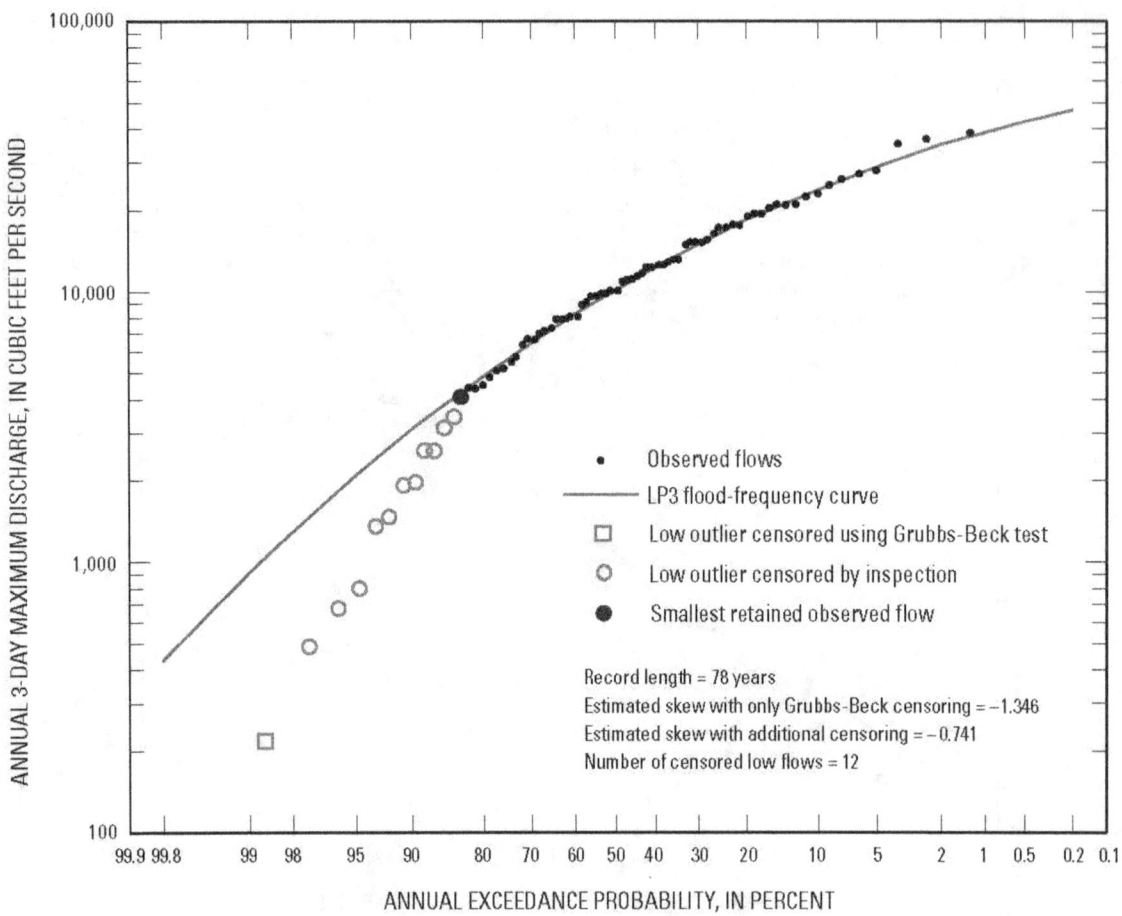

Figure 10. Flood-frequency curve for maximum 3-day duration flows for Putah Creek at Monticello Dam (site 44), California, after additional censoring.

After calculating the skew at each site for each selected *n*-day duration, the data were examined to determine whether the calculated skew showed consistent trends over the selected durations and whether relations between skew and selected basin characteristics were apparent. Figure 11 shows a plot of sample-skew coefficient for each duration on the y-axis versus the sites, ordered by ascending 7-day skew coefficients along the x-axis. Calculated skew coefficients ranged from −1.092 to +0.248. At-site skew coefficients for different durations showed no consistent trends. For example, three sites had a consistently decreasing skew with increasing *n*-day duration, and six sites had a consistently increasing skew with increasing *n*-day duration. At all other sites, skews varied in no consistent way with *n*-day duration. Because basin drainage area is a commonly used characteristic for making hydrologic comparisons, data in figure 11 were replotted in figure 12 against ascending drainage area along the the x-axis. No clear

trend was exhibited between skew coefficients and drainage area, however. In the regional-skew analysis for peak-flow frequency in California, Parrett and others (2011) found that a non-linear function with mean basin elevation represented the site-to-site variability in skew coefficients best. Figure 13 shows the skew coefficient for each duration plotted against the sites ordered in ascending order of mean basin elevation. Skew coefficients for sites with lower mean basin elevations are usually more negative than sites with higher basin elevations. A non-linear trend in skew coefficients for each duration is apparent in figure 13. A transition zone apparently exists between the two clusters of skew values, but the at-site data also exhibit considerable scatter. Appendix 2-3 gives the log-space skew for each site and duration. Statistical analysis of other explanatory variables revealed that only elevation related variables were significant.

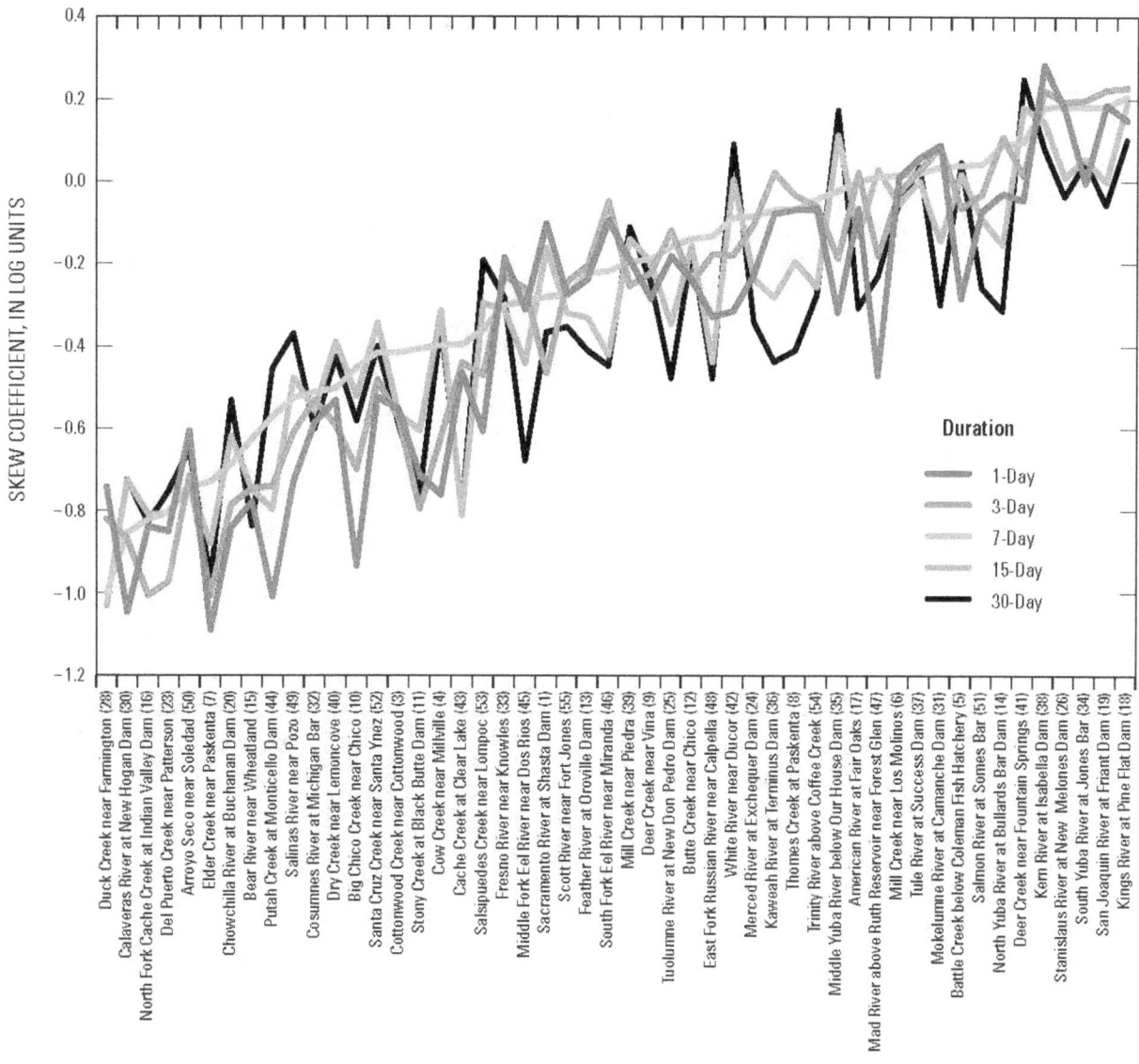

Figure 11. Skew coefficients of rrainfall floods for all durations and for all sites in order of ascending 7-day skew for the study sites in the Central Valley region, California.

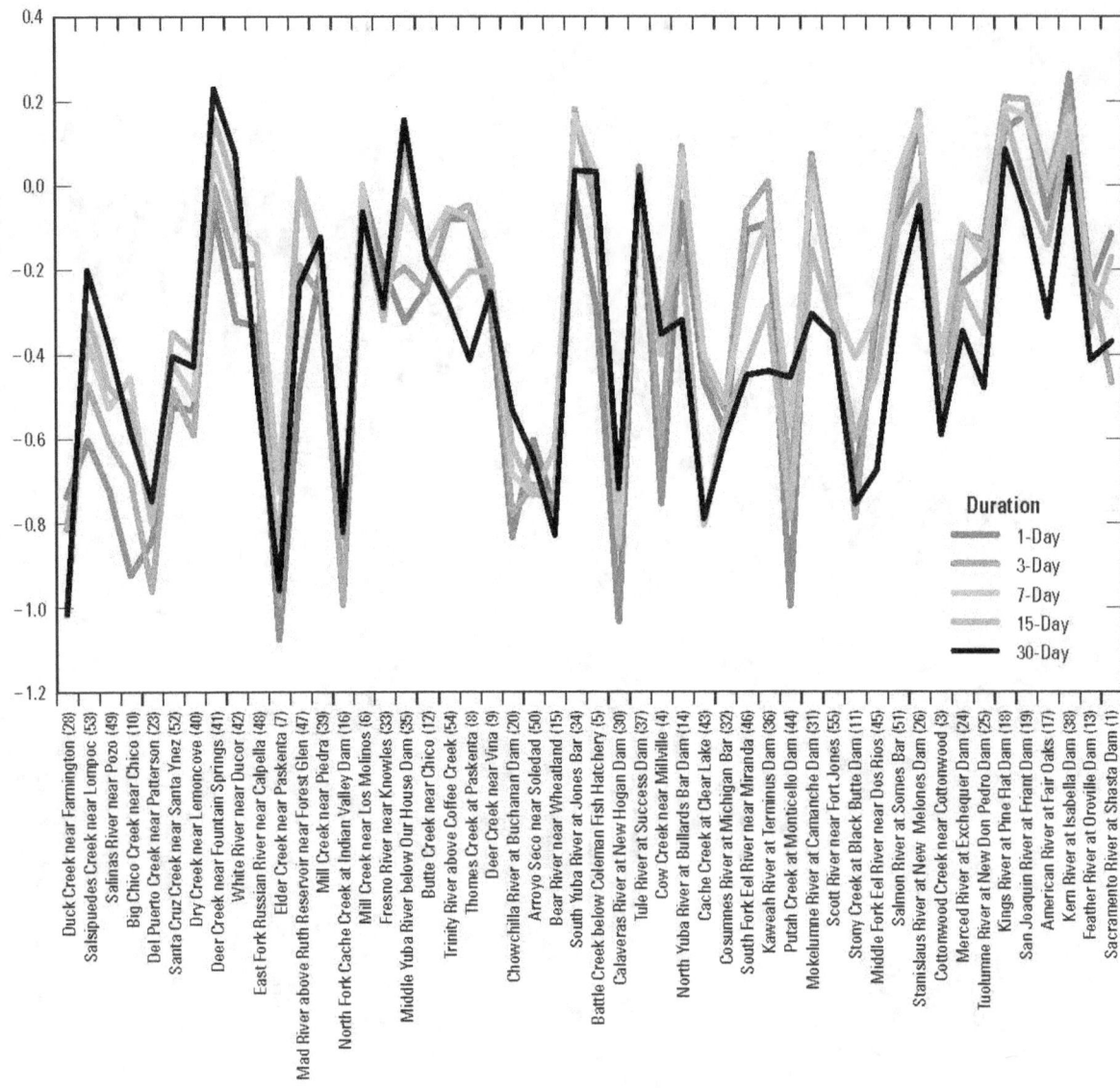

Figure 12. Skew coefficients for 1-, 3-, 7-, 15-, and 30-day rainfall floods ordered by drainage areas in the Central Valley region, California.

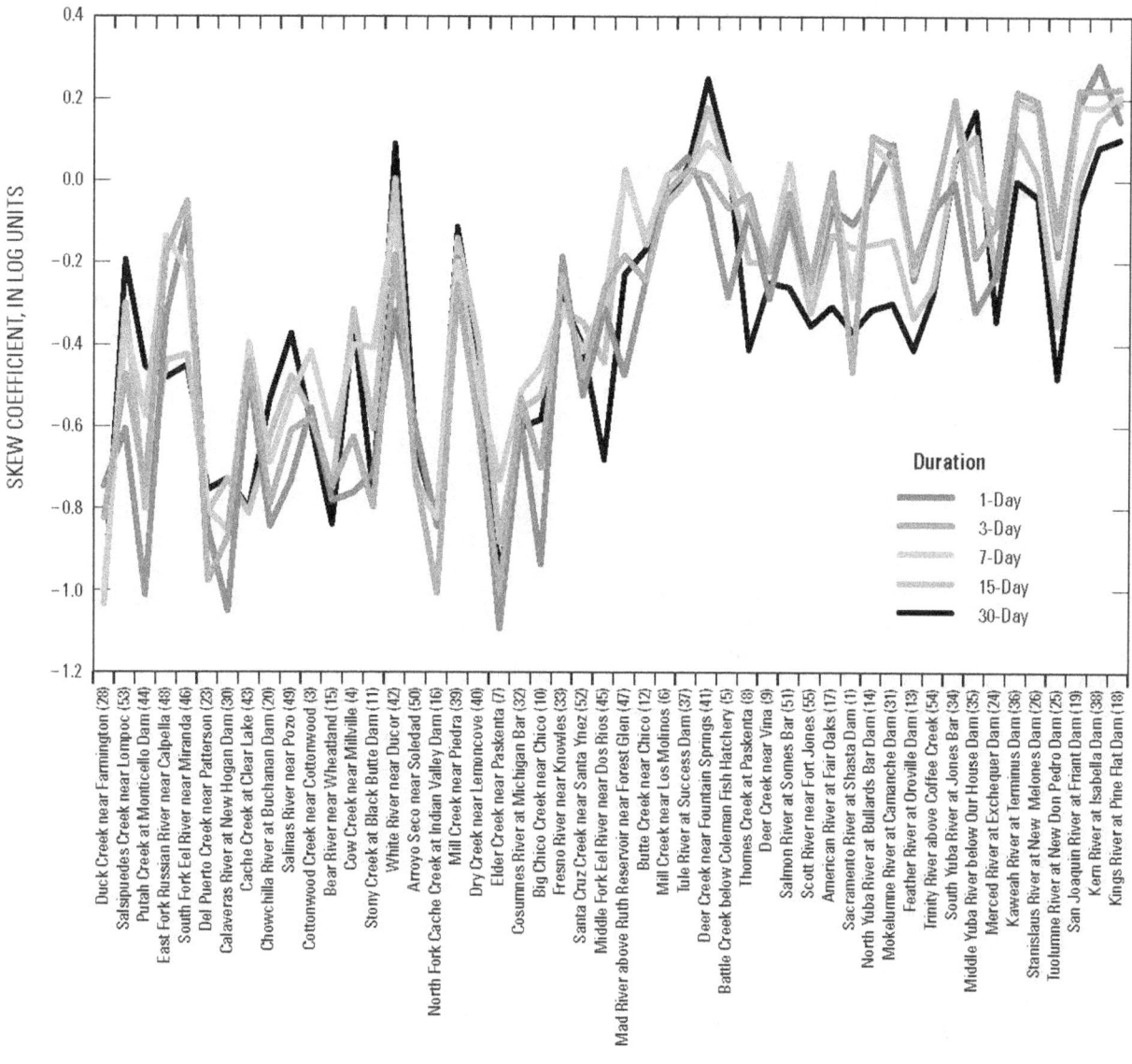

Figure 13.   Skew coefficients for 1-, 3-, 7-, 15-, and 30-day rainfall floods ordered by mean basin elevation in the Central Valley region, California.

Regional Duration-Skew Analysis

Tasker and Stedinger (1986) proposed a weighted least squares (WLS) regression model for estimating regional skew by relating sample skew to basin characteristics. This model accounts for sampling error in the data (a function of record length) as well as model error variance, which describes the precision of the model. Stedinger and Tasker (1985) and Tasker and Stedinger (1989) also presented a generalized least squares (GLS) regression model for estimating flood quantiles from basin characteristics, and Reis and others (2005) presented a Bayesian analysis of that GLS model, which aimed at estimating regional skew. The main advantage of a GLS regression analysis compared to a WLS regression analysis for regional skew is that GLS regression explicitly accounts for sampling error due to cross correlation among skew-coefficient estimators in addition to sampling error due to finite record lengths. This is an important consideration because highly cross-correlated data are not independent. Failure to account for this cross correlation can lead to misrepresentation of model precision. Bayesian GLS regression also is an improvement over traditional GLS regression because Bayesian GLS regression provides the posterior distribution of the model error variance. Moreover, traditional GLS regression can generate a model error variance of zero, which unreasonably indicates that no model error exists (Reis and others, 2005). Bayesian GLS methods have been used to determine regional skew for annual peak flow in the Southeastern U.S. (Feaster and others, 2009; Gotvald and others, 2009; Weaver and others, 2009) and California (Parrett and others, 2011).

A WLS regression was first used to develop the regression model relating regional skew to mean basin elevation, and then a Bayesian GLS model was used to estimate the precision of the WLS regression parameters and the regression diagnostics. For this study, the cross correlations among the annual maximum flows of n-day durations were even greater than those among annual peak flows. Accordingly, a similar hybrid WLS/GLS regression approach to that used by Parrett and others (2011) was used for this study. The hybrid WLS/GLS regression used for this study also used OLS regression to provide initial estimates of the skew coefficients, as described in appendix 3.

Standard GLS Analysis

The GLS model assumes that the regional skew can be described by a linear function of basin and climatic characteristics (explanatory variables) for k explanatory variables and n sites with an additive error (Reis and others, 2005). In matrix notation, the model is as follows:

$$\hat{\gamma} = X\beta + \varepsilon \tag{4}$$

where

$\hat{\gamma}$ is an $(n \times 1)$ vector of unbiased at-site skews for each site;

X is an $(n \times k)$ vector of basin characteristics for each site;

β is a $(k \times 1)$ vector of GLS regression coefficients; and

ε is an $(n \times 1)$ vector of total errors representing the sum of the regional regression model error and the sampling error in the at-site sample-skew estimate for each site.

For this model, $E[\varepsilon] = 0$, and the covariance matrix for ε is $\Lambda = E[\varepsilon\varepsilon^T]$.

The covariance matrix Λ of the vector of errors is given by the equation:

$$\Lambda = \sigma_\delta^2 I + \Sigma(\hat{\gamma}) \tag{5}$$

where

σ_δ^2 is the model error variance, and

$\Sigma(\hat{\gamma})$ is a matrix containing the variances and covariances of the at-site sample-skew estimates for the n site and is a function of record length and cross correlation of annual peaks at different sites.

Given the covariance matrix Λ, the unbiased minimum variance GLS estimator of β, $\hat{\beta}$ is as follows:

$$\hat{\beta} = \left(X^T \Lambda^{-1} X\right)^{-1} X^T \Lambda^{-1} \hat{\gamma} \tag{6}$$

Because Λ is not known, it must be estimated from the data (Reis and others, 2005). The weighted least squares (WLS) estimator of β is obtained when Λ is a diagonal matrix, wherein the cross correlations among the floods at different sites are ignored. For a GLS analysis, the off-diagonal elements of Λ should be the covariance of the sampling errors.

Martins and Stedinger (2002) developed the following relation between the cross correlations of concurrent annual peaks and the cross correlation of at-site skew-coefficient estimators through extensive Monte Carlo experimentation:

$$\hat{\rho}(\hat{\gamma}_i, \hat{\gamma}_j) = \sin(\rho_{ij}) cf_{ij} |\rho_{ij}|^\kappa \qquad (7)$$

where

ρ_{ij} is the cross correlation between concurrent annual peak flows at two gaged sites, i and j,

κ is a constant between 2.8 and 3.3, and

cf_{ij} accounts for the difference between the length of record at each station relative to the concurrent record length and is defined as follows :

$$cf_{ij} = N_{ij} / \sqrt{(N_{ij} + N_i)(N_{ij} + N_j)} \qquad (8)$$

where

N_{ij} is the length of the period of concurrent record, and

N_i, N_j is the number of non-concurrent observations corresponding to i and j.

Thus, given this estimate of the average cross correlation of sample-skew estimates for any two sites and an estimate of their individual sampling variances, the covariance can be calculated.

WLS/GLS Analysis

The standard Bayesian GLS regression approach described previously could not be used in this study because of the high cross correlations among the concurrent flood flows at different sites. Instead, a hybrid WLS/GLS procedure, similar to that developed by Parrett and others (2011) for the regional-skew analysis of peak flows, was developed. This hybrid approach first used OLS regression to estimate at-site skew coefficients for each of the n-day duration flows, which were in turn used to compute the sampling error variance for each of the at-site sample skew coefficient estimates. WLS regression was used secondly to generate robust estimators of the regional-skew model parameters, and GLS regression was used thirdly to estimate the precision of the parameter estimators and the model error variance. The details and mathematics of the hybrid WLS/GLS procedure are described in appendix 3.

Skew-Duration Analysis

In this study, five regional-skew models were generated, one for each of the five selected flood durations (1-day, 3-day, 7-day, 15-day, and 30-day). Skew coefficients for a site should vary modestly from the annual maximum 1-day flood-duration flow to the annual maximum 30-day flood-duration flow. Significant differences in regional skews from one n-day duration to another would mean significant differences in the shape of the LP3 flood-frequency curves and potential inconsistent estimates of n-day flood flows.

This study was based on annual maximum rainfall n-day flood-duration flows at 50 sites in California (fig. 1) having an average of 74 years of record. Because of GIS difficulties and the limited resources available for this study, all basin characteristics for all sites were not available. Basin characteristics other than drainage area, mean basin elevation, relief, maximum elevation, minimum elevation, and basin centroid could not be estimated at three sites. These three sites were site 1, Sacramento River at Shasta Dam; site 14, Feather River at Oroville Dam; and site 20, Chowchilla River at Buchannan Dam. Regression models were first developed and compared by using the data from 47 sites because those sites contained the full set of basin characteristics.

In the WLS/GLS analysis, the estimated variance of the skew for each site depended on the record length available at that site and the skew coefficient for that site estimated from an OLS regression equation relating sample skew to mean basin elevation.

Zero flows and other low outliers were treated as censored observations in the EMA analysis. When the number of censored observations at a site was less than five, the presence of censored observations was ignored when computing the sampling variance estimate for the unbiased skew estimator given by Griffis and Stedinger (2009). On that basis, the only two sites with a greater level of censoring were Santa Cruz Creek near Santa Ynez (site 52) and Putah Creek at Monticello Dam (site 44). For these two sites, the Griffis and Stedinger (2009) expression for the variance of the skew estimator was not appropriate, and the estimate of the sampling error in the skew coefficient produced by the EMA output (PeakfqSA program, Tim Cohn, USGS, written communication, 2010) was adopted, together with an additional factor to reflect the unbiasing of the sample-skew estimator. The unbiasing factor for the skew coefficient described by Tasker and Stedinger (1986) as $(1 + 6/N)$, where N is the length of record, was used with all of the skew estimators.

Of the twenty basin characteristics considered as potential explanatory variables, only mean basin elevation (ELEV) and percentage of basin area with elevation greater than 6,000 ft (EL6000) were found to be statistically significant in the regression analyses. Similarly, mean basin elevation was the only significant basin explanatory variable in the regional-skew regression for annual peak discharge in California (Parrett and others, 2011). Table 5 presents five candidate skew models for each of the n-day flow durations: (1) a model based on a constant value of skew, generally termed a constant model; (2) a model based on a linear relation between skew and mean basin elevation, termed a linear elevation model; (3) a discontinuous constant model with two regression constants, β_0 and β_1, where the value of β_1 is zero for basins where the percentage of area above 6,000 ft elevation (EL6000) is less than 4 percent and equals the value indicated in table 6 for basins where EL6000 is 4 percent or greater; (4) a model based on a nonlinear relation between skew and mean basin elevation; and (5) the same model as 4, but with data based on basin elevation characteristics from all 50 basins, rather than just the 47 basins that had a complete set of basin characteristics. Model 5 (nonlinear equation–final model) thus represents the final and best fit model for regional skew for annual maximum n-day flood flows for all 50 basins.

Table 6 shows several statistics that were used to assess model performance. Pseudo R_δ^2 indicates the fraction of the variation in the true skew that is explained by a model (Gruber and others, 2007). The pseudo R_δ^2 statistic is calculated as follows:

$$R_\delta^2 = 1 - \frac{\sigma_\delta^2(k)}{\sigma_\delta^2(0)} \qquad (9)$$

where

$\sigma_\delta^2(k)$ is model error variance obtained with a model using k explanatory variables, and,

$\sigma_\delta^2(0)$ is model error variance for the constant model, 1.

Another statistic used to assess model performance is the average sample error variance (ASEV). This statistic describes the contribution of the sampling error in the model parameters to the average variance of prediction (Stedinger and Tasker, 1985). ASEV plus the expected model error variance is the

Table 5. Mathematical models used in the regional-skew analyses.

[**Abbreviations**: ELEV, mean basin elevation in feet; EL6000, percentage of basin area above 6,000 feet elevation; γ, regional skew; β_0 and β_1, regional-skew coefficients as defined in table 6; <, less than]

Model type	Regional skew (γ) models
(1) Constant	$\gamma = \beta_0$
(2) Linear elevation	$\gamma = \beta_0 + \beta_1(ELEV)$
(3) Discontinuous EL6000	$\gamma = \beta_0 + \beta_1$, where $\beta_1 = 0$ when EL6000 <4 percent
(4) Nonlinear elevation	$\gamma = \beta_0 + \beta_1[1 - \exp\{-(ELEV/3600)^{12}\}]$
(5) Nonlinear elevation–final	$\gamma = \beta_0 + \beta_1[1 - \exp\{-(ELEV/3600)^{12}\}]$

average value of the variance of prediction for a new site ($E[VP_{new}]$), which is comparable to the MSE reported in Bulletin 17B for the error in the national map of regional skew for annual peak flow. The average ERL is the average effective record length for a regional-skew estimate and indicates the at-site record length required to calculate a skew coefficient with a variance equal to the variance of prediction for the regional-skew model.

As indicated in table 6, the constant model, 1, always has a pseudo R_δ^2 value of zero. In addition, the constant model had larger model error variance (σ_δ^2) and average variance of prediction ($E[VP_{new}]$) than any of the other models for all n-day durations. As a result, the constant model had the shortest average effective record length of any model for all durations. The constant model generally overestimated the skew for basins having a mean elevation of less than about 3,200 ft and generally underestimated the skew for basins having a mean elevation greater than about 4,000 ft.

Although the linear elevation model, 2, had a smaller average model error variance σ_δ^2 than model 1 and a pseudo R_δ^2 value greater than zero, it did not fit the data well for basins with low (less than about 3,000 ft) or high (above about 4,000 ft) mean basin elevations. The Discontinuous constant model, 3, where the value of the constant depends on the value of EL6000, performed better than either model 1 or 2 for all durations on the basis of the regression performance statistics shown in table 6. However, model 3 can provide unrealistic estimates of skew for hydrologically similar and nearby basins if one has an EL6000 just under 4 percent and the other has an EL6000 just over 4 percent. The models based on the nonlinear relation of skew to mean basin elevation (models 4 and 5) represent an attempt to provide a continuous relation of skew to basin elevation over the complete range of both basin elevation and skew, which tends to cluster near a constant value when mean basin elevation is low and at a different constant value when mean basin elevation is high.

Table 6. Summary of statistical results of five regional-skew models for five durations.

[Numbers in parentheses represent the model used and are defined in table 5. Nonsignificant values are **bold**. **Abbreviations**: ASEV, average sample error variance; ERL, effective record length in years; $E[VP_{new}]$, average value of the variance of prediction for a new site; $E[\sigma_\delta^2]$, posterior mean of the model error variance; Pseudo R_δ^2, fraction of the variability in the true skews explained by each model (Gruber and others, 2007); β_0 and β_1, regional skew coefficients; –, not applicable]

Duration (days)	Model type	β_0	β_1	$E[\sigma_\delta^2]$	ASEV	$E[VP_{new}]$	Pseudo R_δ^2	Average ERL
1	(1) Constant [1]	**−0.3197**	–	0.078	0.035	0.113	0	66
	(2) Linear elevation [1]	−1.0235	0.0002	0.026	0.040	0.066	0.665	110
	(3) Discontinuous EL6000 [1]	−0.6903	0.6227	0.017	0.038	0.055	0.780	131
	(4) Nonlinear elevation [1]	−0.7263	0.6923	0.012	0.038	0.049	0.848	146
	(5) Nonlinear elevation – final [2]	−0.7346	0.6859	0.011	0.037	0.048	0.864	150
3	(1) Constant [1]	**−0.2689**	–	0.080	0.039	0.118	0	62
	(2) Linear elevation [1]	−0.9689	0.0002	0.025	0.043	0.068	0.689	104
	(3) Discontinuous EL6000 [1]	−0.6417	0.6290	0.016	0.041	0.057	0.795	122
	(4) Nonlinear elevation [1]	−0.6847	0.7109	0.008	0.040	0.049	0.897	143
	(5) Nonlinear elevation – final [2]	−0.6905	0.6822	0.009	0.040	0.049	0.891	143
7	(1) Constant [1]	**−0.2206**	–	0.053	0.040	0.093	0	76
	(2) Linear elevation [1]	−0.8287	0.0002	0.014	0.045	0.059	0.736	117
	(3) Discontinuous EL6000 [1]	−0.5380	0.5287	0.013	0.043	0.056	0.750	121
	(4) Nonlinear elevation [1]	−0.5812	0.6111	0.007	0.042	0.049	0.873	138
	(5) Nonlinear elevation – final [2]	−0.5877	0.5899	0.007	0.042	0.049	0.875	140
15	(1) Constant [1]	**−0.3027**	–	0.034	0.043	0.076	0	95
	(2) Linear elevation [1]	−0.8802	0.0001	0.010	0.048	0.058	0.713	124
	(3) Discontinuous EL6000 [1]	−0.6017	0.4879	0.008	0.046	0.055	0.752	130
	(4) Nonlinear elevation [1]	−0.6453	0.5685	0.006	0.046	0.052	0.835	138
	(5) Nonlinear elevation – final [2]	−0.6453	0.5493	0.005	0.046	0.051	0.848	141
30	(1) Constant [1]	**−0.3576**	–	0.033	0.044	0.076	0	98
	(2) Linear elevation [1]	−0.8415	0.0001	0.017	0.049	0.066	0.481	113
	(3) Discontinuous EL6000 [1]	−0.6030	0.4044	0.012	0.047	0.059	0.627	125
	(4) Nonlinear elevation [1]	−0.6379	0.4698	0.011	0.047	0.058	0.667	128
	(5) Nonlinear elevation – final [2]	−0.6331	0.4410	0.010	0.046	0.056	0.688	133

[1] Forty-seven sites were used in the regression analysis; three sites (1, 13, and 20) were excluded from the analysis because physiographic data for these sites were not available. See table 1 for a list of sites.

[2] Fifty sites were used in the regression analysis. See table 1 for a list of sites.

The transition in skew from a lesser constant value to a greater constant value was accommodated by using a nonlinear, exponential function of ELEV in the regional regression rather than ELEV itself. After some exploratory trials, the following regression model relating skew and mean basin elevation was developed:

$$\gamma = \beta_0 + \beta_1 \left[1 - \exp\{-(ELEV / 3600)^{12}\} \right] \qquad (10a)$$

or

$$\gamma = \gamma_{min} + (\gamma_{max} - \gamma_{min}) \left[1 - \exp\{-(ELEV / 3600)^{12}\} \right] \qquad (10b)$$

where

ELEV is mean basin elevation in feet,
γ_{min} is the minimum regional-skew coefficient (equal to β_0), and
γ_{max} is the maximum regional-skew coefficient (equal to $\beta_0 + \beta_1$).

Equations 10a and 10b represent two formulations of the same model. Equation 10a expresses regional skew in terms of regression parameters β_0 and β_1, whereas equation 10b expresses regional skew in terms of the maximum and minimum regional-skew values for the study region. This second formulation emphasizes that the regional-skew coefficient model has a minimum skew-coefficient value for low elevation sites and a maximum skew-coefficient value for high elevation sites with a transition occurring around 3,600 ft. The non-linear elevation term (the bracketed portion in both equations) varies between zero at low elevations and one at high elevations. The denominator constant (3,600) inside the exponential function (exp) is a scale parameter that determines the location of the transition between high and low elevation skew coefficients. The exponent (12) inside the exponential function is a shape parameter that controls how rapid the transition is between low and high elevation constant skew coefficients. As indicated by the results in table 6, the nonlinear elevation models (models 4 and 5) provided the best regression fits to the data for all durations on the basis of the regression performance statistics. Model 5, which represents the final, best nonlinear elevation model for regional skew based on data for all 50 basins, generally had slightly better performance statistics than those for model 4, which was based on data from 47 of the 50 sites.

Table 6 shows that the $E[VP_{new}]$ for model 5 ranged from 0.048 for a 1-day flood flow to 0.056 for a 30-day flood flow. The average ERL values for model 5 in table 6 ranged from 133 years for the 30-day flood flow to 150 years for the 1-day flood flow. In contrast, the mean squared error of 0.303 reported for the National skew map in Bulletin 17B corresponds to only 17 years of effective record length for the estimate of the skew coefficient based on annual peak flows. Overall, model errors for model 5 are notably small, the fitted functions are reasonable, and the equivalent years of record are notably long.

Results from the pseudo analysis of variance (ANOVA) for model 5 for each duration are given in table 7. The analysis divides the variability observed in the skew estimators into three sources: the variability explained by the model, the variability in the true skew that the model cannot explain (model error), and the variability due to the sampling error in the individual skew estimators. The model error describes the precision with which the regression model can estimate the "true" skew. For all durations, the model error was much less than the sampling error. The table also reports the total variability, which is the sum of the three variabilities (model, model error, and sampling error). The major source of variability for all durations is the sampling error.

The error variance ratio (EVR) in table 7 is the average sampling variance divided by the variance of the model error. This statistic is used to determine if an OLS analysis is adequate or if a WLS or GLS analysis is needed. Values less than about 0.2 indicate that an OLS analysis is appropriate, whereas values much greater than one indicate a WLS or GLS analysis is needed. EVR values ranged from 12.4 to 26.3 across all durations, indicating that an WLS or GLS analysis was needed.

Table 7. Pseudo ANOVA table for the final non-linear regional-skew model for all *n*-day flood durations, Central Valley region, California.

[**Abbreviations**: ANOVA, analysis of variance; EVR, error variance ratio; MBV, misrepresentation of the beta variance]

Source	Durations				
	1-Day	3-Day	7-Day	15-Day	30-Day
Model	3.384	3.660	2.347	1.359	1.070
Model error	0.533	0.448	0.336	0.244	0.485
Sampling error	6.602	6.439	6.234	6.399	6.348
Total	10.519	10.548	8.916	8.002	7.902
EVR	12.4	14.4	18.6	26.3	13.1
MBV	13.4	15.2	17.1	18.4	18.0
Pseudo R^2	0.86	0.89	0.87	0.85	0.69

Table 7 also reports the misrepresentation of beta variance (MBV) statistic. MBV is the ratio of the sampling variance that the GLS regression analysis ascribes to the constant in the model to the variance that a WLS regression analysis (that neglects cross correlations) would ascribe to the constant (Parrett and others, 2011). The MBV statistic is used to determine if a WLS regression analysis of model precision is adequate or if a GLS regression analysis is required. If MBV values are much greater than one, a GLS analysis is needed to properly assess model precision. MBV values ranged from 13.4 to 18.4 across all durations, indicating that the error analysis produced by a WLS regression analysis would overestimate the precision of the constant term. Thus, a GLS regression analysis is needed to correctly evaluate the precision with which the constant term can be resolved. This is particularly important for these analyses because the contribution of parameter uncertainty to the average variance of prediction is at least twice as great as the model error variance. The sampling error is usually the predominant source of error in regional skew-coefficient predictions which is caused by the large correlations among flood records that limit the amount of information a regional dataset provides.

Figure 14 shows the at-site sample-skew coefficients plotted against their mean basin elevations together with the fitted curve from model 5 for 1-day, 3-day, 7-day, 15-day, and 30-day durations. There is much scatter in the data displayed in these figures, largely because of the sampling error in the skew-coefficient estimators. Moreover, residual errors were also correlated because of the many correlations among the at-site annual maximum flood flows for each duration. Despite the considerable scatter in the at-site data in figure 14, the data and the model curves indicated a significant change in skew as mean basin elevation increased from about 3,000 ft to about 4,000 ft. The changes in skew with increasing mean basin elevation were more dramatic for annual peak flows (Parrett and others, 2011) probably because the regional-skew analysis for annual peak discharge used all (snowmelt- and rainfall-caused) annual peak-flow data and not just annual peak flows from rainfall. Nevertheless, the significant, albeit smaller, changes in skew with increasing elevation in this study indicated that increasing basin elevation changes flood response even when annual maximum flood flows from snowmelt were eliminated from the data. The changes in flood response are still probably largely related to increasing snow effects at higher elevations. Thus, some storms that are rainfall at lower elevations can be in the form of snow at higher elevations. In addition, runoff from warm rain can be intensified by snowmelt at high elevations.

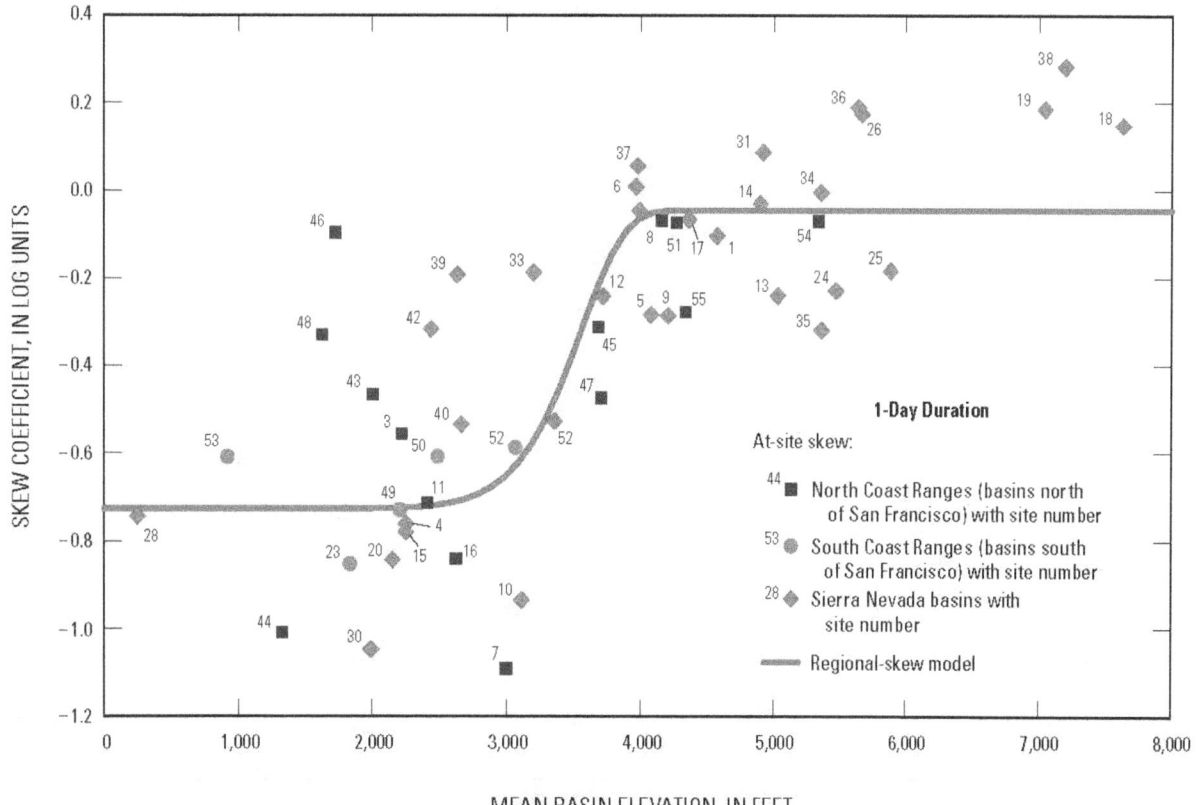

Figure 14. A function of mean basin elevation with the regional model for durations of (A) 1-day (B) 3-days, (C) 7-days, (D) 15-days, and (E) 30-days.

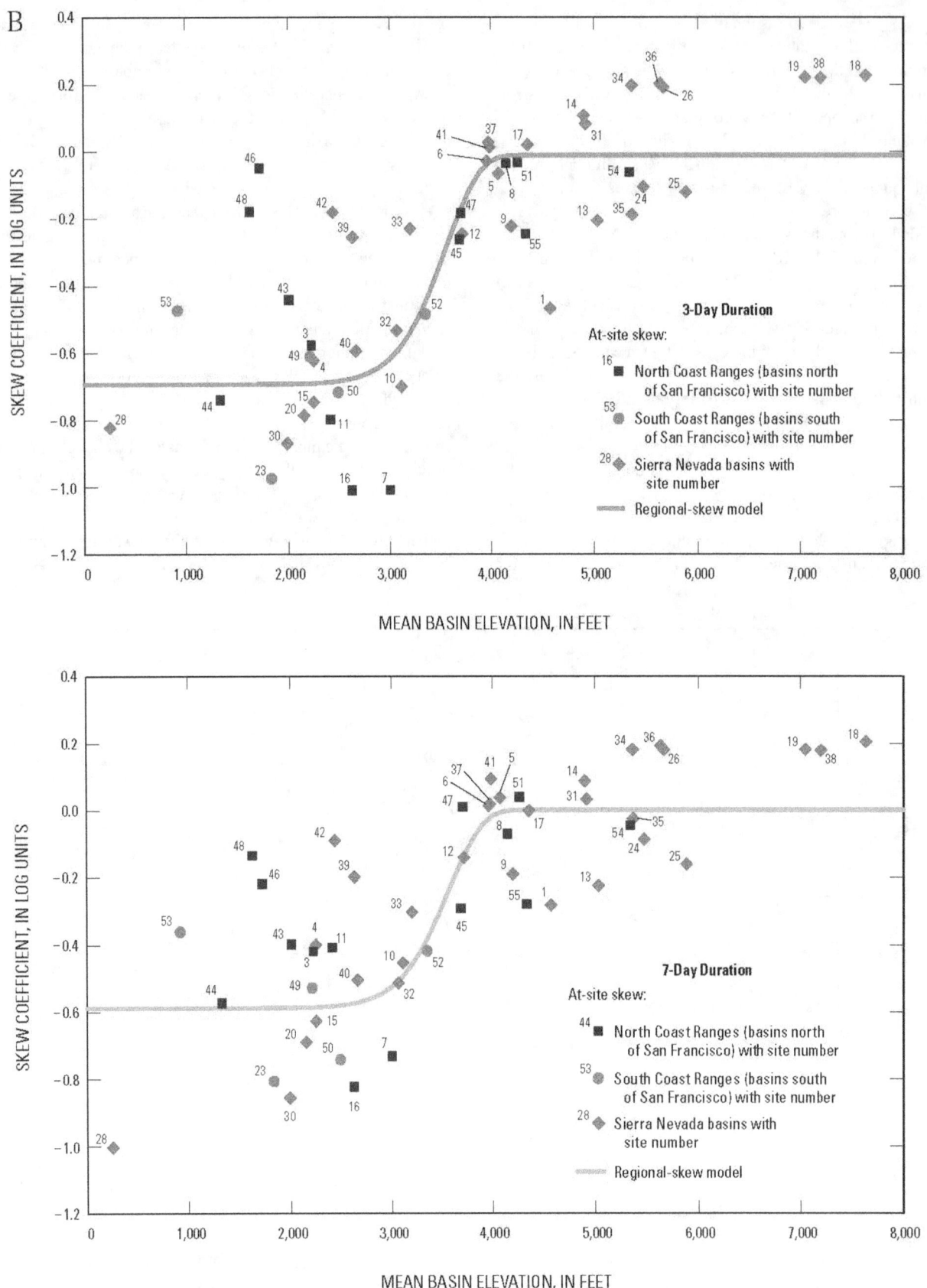

Figure 14.—Continued

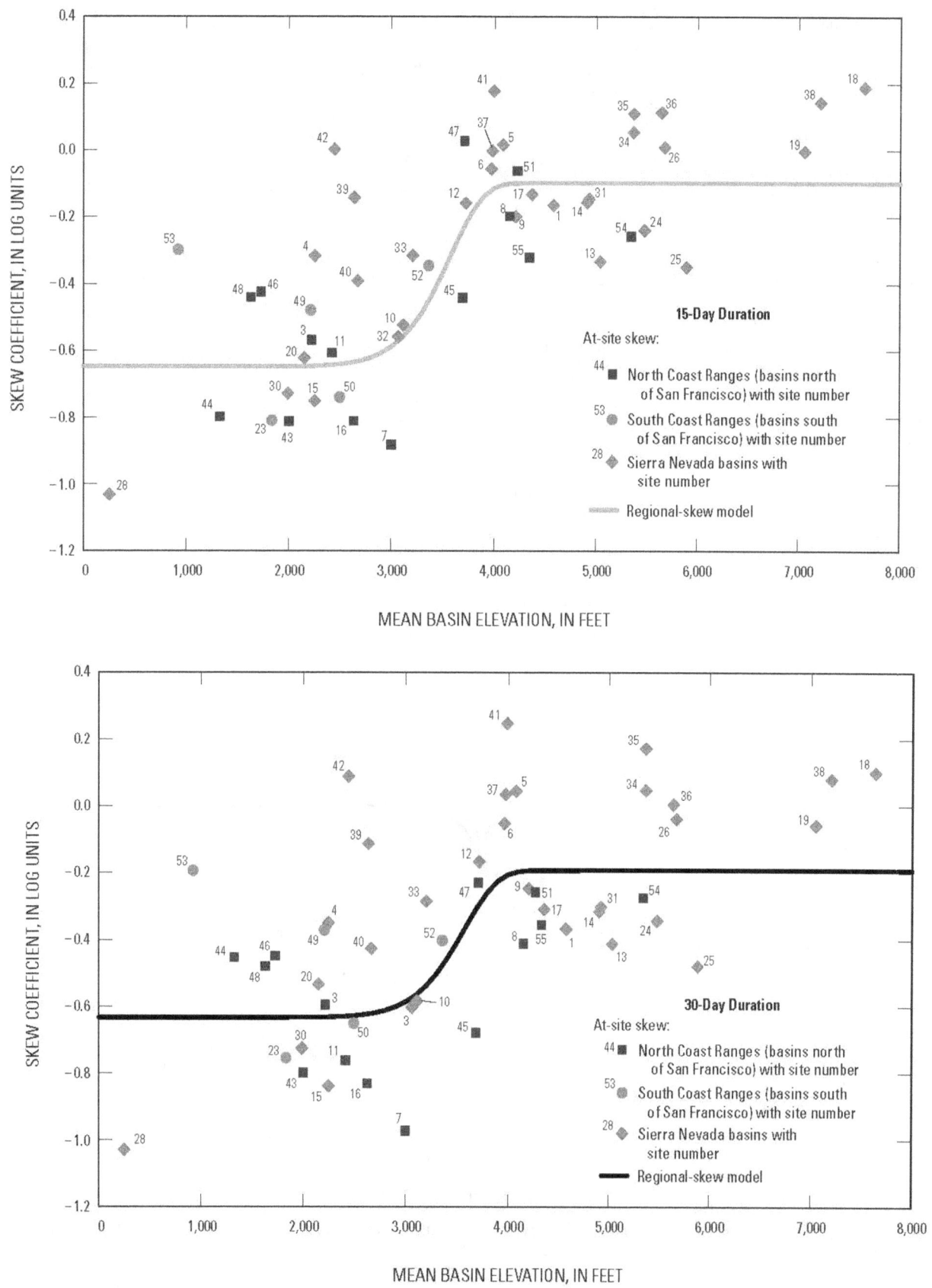

Figure 14.—Continued

The regional-skew models for each duration developed from the final model, 5, are represented by the curves in figure 15. Overall, regional skew ranged from −0.74 for the 1-day duration flood flow for mean basin elevations less than 2,500 ft to about 0 for the 7-day duration flood flow for mean basin elevations greater than 4,500 ft. The difference between the minimum and maximum skew coefficients was greatest for the 1-day duration flood model and least for the 30-day duration flood model. The differences between maximum and minimum skew were somewhat less for the longer duration flood flows (15- and 30-day duration) than for the shorter duration flood flows (1- and 3-day duration). These differences, though subtle, could indicate that flood-frequency characteristics are more uniform across the study region for longer duration floods than for shorter duration floods. The dampening effect of averaging daily maximum flows over longer time spans for the longer durations probably is partly responsible for the increased uniformity. Another possible explanation is that basin infiltration becomes significantly reduced as soils become saturated after prolonged storms. The effects of variable infiltration characteristics on runoff thus become reduced for longer duration flood flows. Overall, the results in table 6 indicate that the 30-day duration

model has somewhat more scatter (lowest psuedo R^2 and shortest effective record length) and a lower elevation signal (smaller beta coefficient for mean basin elevation) than any duration model.

Because the models include an explanatory variable that depends on elevation, the actual variance of prediction for a site depends on its mean basin elevation (ELEV). Table 8 gives the variance of prediction for a site not included in this study (VP_{new}) as a function of its mean basin elevation between 0 and 10,000 ft. For sites with mean basin elevations less than 2,500 ft, skew was constant, and the variance of prediction did not change with mean basin elevation. Similarly, sites with mean basin elevations greater than 4,500 ft had a constant skew, and the variance of prediction did not change with mean basin elevation. The variation in the effective record lengths (ERL) was low with respect to changes in mean basin elevation despite an appreciable variation in the variance of prediction. The change in the sampling variance of the skew estimators due to the change in estimated skew with elevation was approximately balanced by the difference in the prediction variance between lower and higher elevation basins.

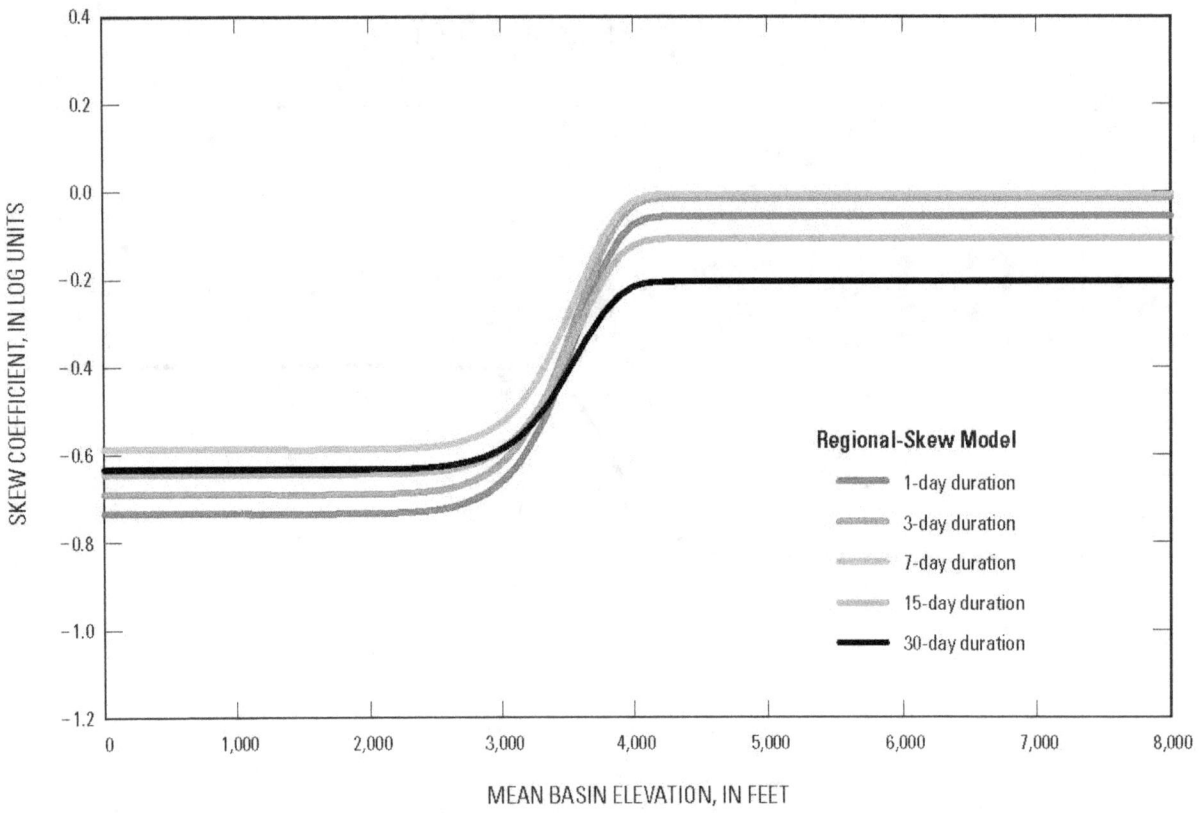

Figure 15.   Models of nonlinear skews for all durations in the Central Valley region, California.

Table 8.   Variance of Prediction (VP) and Effective Record Length (ERL) for five durations as a function of mean basin elevation.

[**Abbreviations**: >, greater than; <, less than]

Mean basin elevation (ELEV) (feet)	Durations									
	1-day		3-day		7-day		15-day		30-day	
	VP_{new}	ERL (years)	VP_{new}	ERL (years)	VP_{new}	ERL (years)	VP_{new}	ERL (years)	VP_{new}	ERL (years)
<2,500	0.058	186	0.059	172	0.058	156	0.062	157	0.066	145
3,000	0.055	182	0.056	168	0.055	155	0.059	156	0.063	144
3,200	0.052	177	0.053	164	0.053	153	0.055	155	0.060	144
3,400	0.047	170	0.049	159	0.049	151	0.051	154	0.056	143
3,600	0.043	164	0.044	155	0.045	151	0.046	154	0.052	142
3,800	0.040	162	0.042	155	0.042	153	0.042	156	0.049	141
4,000	0.039	162	0.041	157	0.041	156	0.041	157	0.048	141
>4,500	0.039	162	0.040	157	0.041	156	0.041	157	0.048	140

Given the high degree of censoring for some sites, there was concern that large sampling variance at those sites could adversely affect the statistical analysis. As a check, the WLS/GLS regression analysis for the nonlinear elevation–final model (model 5) for each duration was rerun without the four most heavily censored sites (site 43, Cache Creek at Clear Lake; site 16, North Fork Cache Creek at Indian Valley Dam; site 52, Santa Cruz Creek near Santa Ynez; and site 44, Putah Creek at Monticello Dam). Differences in model results and regression diagnostics were minimal, indicating that censoring at these four sites had little effect on the final model results. Appendix 2-4 shows the regional-skew coefficient estimates for the 50 study sites at the five durations. The associated variance of prediction for weighting the regional sample skew-coefficient estimator is given in appendix 2-5.

Use of Regional-Skew Models

The previous section showed that regional skews are more accurate, and mean squared errors (MSE) are less than in the National skew map of Bulletin 17B. Regional skew and MSE, along with the corresponding at-site values, are needed to estimate a weighted skew coefficient necessary for flood-frequency estimates at gaged sites. This is the case because the shape, or skew, of the flood distribution is often significantly affected by the presence of very small or very large discharges in the record (outliers) and also by the length of the record.

The non-linear regional-skew models for each duration (1-day, 3-day, 7-day, 15-day, and 30-day) are presented in figures 14 and 15. These figures, or equation 10a, can be used to estimate the regional skew at a gaged site, but the mean basin elevation for a site must be known. For example, the regional-skew coefficient for a 3-day duration is needed for a site with a mean basin elevation (ELEV) of 3,500 ft. In figure 14B, the regional-skew coefficient is approximately −0.34, the intersection of 3,500 ft and the model (green line). Using equation 10a with an elevation of 3,500 ft, a regional-skew coefficient was calculated to be −0.3426 with the values of β_0 (-0.6905) and β_1 (0.6822) given from table 6. For sites used in this study, the regional skews for each duration are given in appendix 2-4.

The variance of prediction (VP) is equivalent to the MSE from the National map of regional skew for annual peak flow of Bulletin 17B. Table 6 reports the VP as $E[VP_{new}]$, which is an average for each model and duration. Because skew is a function of elevation, table 8 presents the VP for each duration as a function of ELEV (mean basin elevation). From the previous example, the MSE for a 3-day duration regional skew is 0.0465, which was calculated by averaging the VP values surrounding 3,500 ft (0.0465 = (0.049 + 0.044)/2) from table 8. For sites used in this study, the MSEs (which is equivalent to VP_{old}) for each duration are given in appendix 2-5.

Summary

Accurate and reliable estimates of the magnitude and frequency of flood flow volumes for a *n*-day duration are critical for the evaluation of the risk of flooding and the operation and reliability of dams and levees. Recognizing the need for accurate estimates of volume-duration frequencies in the Central Valley region of California, the U.S. Geological Survey, in cooperation with the U.S. Army Corps of Engineers (USACE), conducted a study to develop regional-skew models for the 1-day, 3-day, 7-day, 15-day, and 30-day rainfall-flood durations. The analysis of regional-skew coefficient estimators is important because flood-frequency estimates can be determined with greater accuracy by using the more precise estimates of the skew coefficients. This report documents the methods in developing regional-skew models for five flood-flow durations for the Central Valley region. Fifty sites, all but four of which had records through either 2008 or 2009, were used in the development of regional-skew coefficient models. Twenty-two of these sites were at dams, and the daily unregulated-flow records at these sites were synthesized from records of flow, reservoir storage levels, and diversions. The other 28 sites had no significant regulation during the study period. The record at one site was extended by using the MOVE.1 technique with flow records from several other sites.

Flood-frequency analysis is usually conducted on peak flows, but peak flows are not as critical to the operations of large dams and reservoirs as much as sustained flows (volume-duration). The 3-day maximum rainfall flood volume is the most critical duration found by USACE and the Bureau of Reclamation for much of the Central Valley region of California because of the many large control structures. The 7-day maximum rainfall flood volume is also important because it can represent two 3-day back-to-back events which is not an uncommon meteorological event in the region. In accordance with recommendations in Bulletin 17B, the Pearson Type III distribution applied to the logarithms (base 10) of the selected annual maximum rainfall flood-duration data was used to determine flood-frequency statistics at each site for this study. The expected moment algorithm (EMA) was used, when necessary, for fitting the LP3 distribution in order to determine station skew for sites used in the regional-skew analysis that had flood-duration flows identified as low outliers and (or) zero flows.

This study employed recently developed generalized least squares (GLS) regression procedures for regional skew analyses. To properly account for the high cross correlations among annual peak discharges and in the skew-coefficient estimators, a combination of Bayesian weighted least squares (B-WLS) and Bayesian generalized least squares (B-GLS) regression was adopted to ensure that the regression model and the diagnostics for the regression analysis were reliable.

Several basin characteristics were considered as possible explanatory variables in the regression analysis for regional skew. The basin characteristic that explained the site-to-site variability in skew best was the mean basin elevation. Five skew models were developed: (1) a model that uses a constant skew, (2) a model that uses a linear relation between skew and mean basin elevation , (3) a discontinuous model that uses one constant term for sites with EL6000 less than or equal to 4 percent and another for sites with EL6000 greater than 4 percent, (4) a model that uses a nonlinear relation between skew and mean basin elevation, and (5) a model that uses the same nonlinear relation between skew and mean basin elevation as model 4, but is based on data from 50 sites rather than the 47 sites used to develop models 1 through 4. The nonlinear elevation–final model (model 5) provided a reasonable fit to the data and had smaller model errors and greater pseudo R_d^2 values than the other models. The average value of the variance of prediction at a new site ($E[VP_{new}]$) corresponds to the mean square error (MSE) for the regional-skew estimator. It describes the precision of the generalized skew. $E[VP_{new}]$ was smallest for the final model, 5. Just as the generalized skew coefficient varies from site-to-site depending on mean basin elevation, so too does the value of the variance of prediction for a new site, VP_{new}. The final regional-skew model, 5 had VP_{new} values ranging from about 0.041 to 0.066 and a corresponding effective record length (ERL) between 140 years and 186 years, depending upon the values of mean basin elevation and flood duration. In contrast, the National skew map for peak flows of Bulletin 17B has a MSE of 0.302 and ERL of only 17 years.

References

Cohn, T.A., Lane, W.L., and Baier, W.G., 1997, An algorithm for computing moments-based flood quantile estimates when historical flood information is available: Water Resources Research, v. 39, no. 9, p. 2089–2096.

Cohn, T.A., Lane, W.L., and Stedinger, J.R., 2001, Confidence intervals for EMA flood quantile estimates: Water Resources Research, v. 37, no. 6, p. 1695–1706.

Cudworth, A.G. Jr., 1989, Flood Hydrology Manual: U.S. Department of the Interior, Bureau of Reclamation, A Water Resources Technical Publication, 243 p.

England, J.F. Jr., Jarrett, R.D., and Salas, J.D., 2003b, Data-based comparisons of moments estimators that use historical and paleoflood data: Journal of Hydrology, v. 278, no. 1–4, p. 170–194.

England, J.F. Jr., Salas, J.D., and Jarrett, R.D., 2003a, Comparisons of two moments-based estimators that utilize historical and paleoflood data for the log-Pearson type III distribution: Water Resources Research, v. 39, no. 9, 1243, 16 p., doi:10.1029/2002WR001791.

Feaster, T.D., Gotvald, A.J., and Weaver, J.C., 2009, Magnitude and frequency of rural floods in the southeastern United States, 2006—Volume 3, South Carolina: U.S. Geological Survey Scientific Investigations Report 2009–5156, 226 p.

Gotvald, A.J., Feaster, T.D., and Weaver, J.C., 2009, Magnitude and frequency of rural floods in the southeastern United States, 2006—Volume 1, Georgia: U.S. Geological Survey Scientific Investigations Report 2009–5043, 120 p.

Griffis, V.W., and Stedinger, J. R., 2009, Log-Pearson type 3 distribution and its application in flood frequency analysis, III: sample skew and weighted skew estimators: Journal of Hydrology, v. 14, no. 2, p. 121–130.

Griffis, V.W., Stedinger, J.R., and Cohn, T.A., 2004, LP3 quantile estimators with regional skew information and low outlier adjustments: Water Resources Research, v. 40, W07503, doi:1029/2003WR002697.

Gruber, A.M., Reis, D.S., Jr., and Stedinger, J.R., 2007, Models of regional skew based on Bayesian GLS regression, Paper 40927-3285, in K.C. Kabbes ed., Restoring our Natural Habitat: Proceedings of the World Environmental and Water Resources Congress 2007, Tampa, Florida, American Society of Civil Engineers, May 15–18, 2007.

Gruber, A.M., and Stedinger, J.R., 2008, Models of LP3 regional skew, data selection, and Bayesian GLS regression, in Babcock, Jr., R.W., and Walton, Raymond, eds., Ahupua'a: Proceedings of the World Environmental and Water Resources Congress 2008, paper 596, Honolulu, Hawaii, American Society of Civil Engineers, May 12–16, 2008.

Helsel, D.R., and Hirsch, R.M., 1992, Statistical methods in water resources: Amsterdam, the Netherlands, Elsevier Science Publishers, 522 p.

Hickey, J.T., Collins, R.F., High, J.M., Richardsen, K.A., White, L.L, and Pugner, P.E., 2002, Synthetic Rain Flood Hydrology for the Sacramento and San Joaquin River Basins: Journal of Hydrologic Engineering, v. 7, Issue 3, p. 195–208.

Interagency Advisory Committee on Water Data, 1982, Guidelines for determining flood-flow frequency, Bulletin #17B of the Hydrology Subcommittee, Office of Water Data Coordination: U.S. Geological Survey, Reston Virginia, 183 p. Available at http://water.usgs.gov/osw/bulletin17b/dl_flow.pdf.

Martins, E.S., and Stedinger, J.R., 2002, Cross-correlation among estimators of shape: Water Resources Research, v. 38, no. 11, 1252, 7p., doi:10.1029/2002WR001589

Mount, J.F., 1995, California rivers and streams the conflict between fluvial process and land use, University of California Press, p. 94–100.

National Research Council, 1999, Improving American River flood frequency analyses: National Academies Press, Washington, D.C., 120 p.

Parrett, C., Veilleux, A., Stedinger, J.R., Barth, N.A., Knifong, D.L., and Ferris, J.C., 2011, Regional skew for California, and flood frequency for selected sites in the Sacramento-San Joaquin River Basin, based on data through water year 2006: U.S. Geological Survey Scientific Investigations Report 2010–5260, 94 p.

Reis, D.S., Jr., Stedinger, J.R., and Martins, E.S., 2005, Bayesian generalized least squares regression with application to the log Pearson type III regional skew estimation: Water Resources Research, v. 41, W10419, doi:10.1029/2004WR003445.

Stedinger, J.R., and Tasker, G.D., 1985, Regional hydrologic analysis, 1, ordinary, weighted and generalized least squares compared: Water Resources Research, v. 21, no. 9, p. 1421–1432 [with correction, Water Resources Research, 1986, v. 22, no. 5, p. 844.]

Tasker, G.D., and Stedinger, J.R., 1986, Regional skew with weighted LS regression: Journal of Water Resources Planning and Management, American Society of Civil Engineers, v.112, no. 2, p. 225–237.

Tasker, G.D., and Stedinger, J.R., 1989. An operational GLS model for hydrologic regression: Journal of Hydrology, v. 111, no. 1–4, p. 361–375.

U.S. Army Corps of Engineers, Sacramento District, 1997, Engineering and Design—Hydrologic for Reservoirs: U.S. Army Corps of Engineers, Engineer Manual 1110-2-1420, [p. 115], at http://140.194.76.129/publications/eng-manuals/em1110-2-1420/toc.htm.

U.S. Army Corps of Engineers, Sacramento District, 2002, Technical studies documentation: Sacramento and San Joaquin River Basins comprehensive study, Main report and appendices A through G: U.S. Army Corps of Engineers, accessed July 1, 2011, at http://www.spk.usace.army.mil/projects/civil/compstudy/reports/html.

Weaver, J.C., Feaster, T.D., and Gotvald, A.J., 2009, Magnitude and frequency of rural floods in the southeastern United States, 2006—Volume 2, North Carolina: U.S. Geological Survey Scientific Investigations Report 2009–5158, 113 p.

Appendix 1. Unregulated Annual Maximum Rain Flood Flows for Selected Durations for all 50 Sites in the Central Valley Region Study Area, California.

[The flood-duration data in this appendix were provided by U.S. Army Corps of Engineers. The n-day flood-duration flow is the maximum average discharge of any consecutive n-day period in a water year for a site]

This appendix is available for download at http://pubs.usgs.gov/2012/5130/Appendix1.

Appendix 2. Ancillary Tables for Regional-Skew Study in the Central Valley Region of California

Appendix 2-1. Basin characteristics for all 50 sites included in this study of the Central Valley region of California.

[Site locations are shown in figure 1. See table 2 for definition of basin characteristics. **Abbreviations:** F, Fahrenheit; –, could not be determined]

Site number	Site name	ELEV (feet)	BASINPERIM (miles)	RELIEF (feet)	DRNAREA (square miles)	ELEVMAX (feet)	MINBELEV (feet)	LAKEAREA (percent)	ELG000 (feet)	OUTLETELEV (feet)	RELRELF (feet per mile)	CENTROIDX (decimal degrees)
1	Sacramento River at Shasta Dam	4,571	–	13,147	6,403	14,120	974	0.09	1.66	366	–	-121.364
3	Cottonwood Creek near Cottonwood	2,221	200	7,712	922	8,075	363	0.12	1.88	385	38.66	-122.701
4	Cow Creek near Millville	2,251	140	6,671	423	7,054	383	0.09	15.36	420	47.65	-121.990
5	Battle Creek below Coleman Fish Hatchery	4,074	139	9,953	361	10,371	418	0.04	13.77	385	71.41	-121.749
6	Mill Creek near Los Molinos	3,962	143	9,918	131	10,303	385	0.01	3.20	728	69.38	-121.673
7	Elder Creek near Paskenta	2,998	65	5,920	93	6,647	727	0.25	9.93	699	90.88	-122.659
8	Thomes Creek at Paskenta	4,146	106	7,378	204	8,076	698	0.01	6.19	500	69.75	-122.783
9	Deer Creek near Vina	4,199	141	7,357	209	7,853	496	0.00	0.01	301	52.17	-121.574
10	Big Chico Creek near Chico	3,111	93	5,727	72	6,019	293	1.33	1.63	493	61.81	-121.647
11	Stony Creek at Black Butte Dam	2,416	251	6,611	740	7,049	439	0.27	10.78	327	26.34	-122.619
12	Butte Creek near Chico	3,717	124	6,851	148	7,178	327	0.00	0.00	327	55.30	-121.563
13	Feather River at Oroville Dam	5,031	–	8,546	3,591	9,260	714	–	–	–	–	-120.894
14	North Yuba River at Bullards Bar Dam	4,899	175	7,117	489	8,502	1,384	1.24	26.92	1,387	40.69	-120.850
15	Bear River near Wheatland	2,250	178	5,734	292	5,812	79	1.56	32.17	89	32.17	-121.020
16	North Fork Cache Creek at Indian Valley Dam	2,627	84	4,487	120	5,945	1,458	2.67	0.00	1,474	53.42	-122.643
17	American River at Fair Oaks	4,356	362	10,181	1,887	10,363	182	1.57	26.66	215	28.11	-120.583
18	Kings River at Pine Flat Dam	7,634	275	13,660	1,544	14,226	566	1.22	72.21	567	49.60	-118.337
19	San Joaquin River at Friant Dam	7,046	324	13,463	1,639	13,821	358	1.85	66.07	399	41.49	-119.209
20	Chowchilla River at Buchanan Dam	2,152	–	6,025	235	6,443	418	–	–	–	–	-119.859
23	Del Puerto Creek near Patterson	1,835	66	3,461	73	3,664	203	0.00	0.00	209	52.77	-121.357
24	Merced River at Exchequer Dam	5,473	268	12,658	1,038	13,058	400	0.89	44.80	400	47.25	-119.760
25	Tuolumne River at New Don Pedro Dam	5,882	344	12,723	1,533	13,039	317	2.03	48.22	317	36.95	-119.872
26	Stanislaus River at New Melones Dam	5,663	244	10,811	904	11,544	733	2.29	50.02	1,090	44.25	-120.077
28	Duck Creek near Farmington	249	30	328	11	456	127	0.28	0.00	128	10.96	-120.904
30	Calaveras River at New Hogan Dam	1,991	147	5,549	372	6,071	522	1.23	0.00	522	37.85	-120.589
31	Mokelumne River at Camanche Dam	4,918	284	10,278	628	10,369	90	2.56	37.11	92	36.15	-120.324
32	Cosumnes River at Michigan Bar	3,064	175	7,548	535	7,728	180	0.23	5.41	188	43.08	-120.633
33	Fresno River near Knowles	3,201	91	6,045	134	7,144	1,098	0.03	1.53	1,100	66.10	-119.682
34	South Yuba River at Jones Bar	5,362	161	7,978	311	9,048	1,070	1.80	44.27	1,070	49.67	-120.668
35	Middle Yuba River below Our House Dam	5,365	109	6,365	145	8,363	1,998	1.04	40.51	2,008	58.58	-120.704
36	Kaweah River at Terminus Dam	5,635	161	12,058	560	12,570	513	0.51	45.38	515	74.86	-118.782
37	Tule River at Success Dam	3,975	131	9,575	392	10,224	649	0.48	21.59	655	73.35	-118.739
38	Kern River at Isabella Dam	7,198	370	11,996	2,075	14,469	2,473	0.67	69.20	2,575	32.42	-118.321
39	Mill Creek near Piedra	2,637	87	6,637	115	7,201	564	0.01	0.63	564	76.04	-119.156
40	Dry Creek near Lemoncove	2,668	70	6,992	76	7,559	567	0.00	0.95	568	99.64	-119.004
41	Deer Creek near Fountain Springs	3,989	58	7,231	83	8,226	995	0.00	11.98	995	124.88	-118.690
42	White River near Ducor	2,443	74	7,551	91	8,266	715	0.00	2.08	719	101.52	-118.780
43	Cache Creek at Clear Lake	2,004	184	3,508	527	4,816	1,308	11.84	0.00	1,309	19.03	-122.840
44	Putah Creek at Monticello Dam	1,327	191	4,508	567	4,717	209	5.00	0.00	265	23.56	-122.399
45	Middle Fork Eel River near Dos Rios	3,685	222	6,659	745	7,570	911	0.30	5.40	912	30.05	-123.074
46	South Fork Eel River near Miranda	1,726	186	3,992	537	4,215	224	0.20	0.00	224	21.44	-123.704
47	Mad River above Ruth Reservoir near Forest Glen	3,705	62	3,332	94	6,022	2,690	0.05	0.02	2,690	53.46	-123.244
48	East Fork Russian River near Calpella	1,630	65	3,104	92	3,913	809	0.04	0.00	841	47.88	-123.095
49	Salinas River near Pozo	2,211	56	2,697	70	4,058	1,361	0.01	0.00	1,362	47.84	-120.298
50	Arroyo Seco near Soledad	2,494	110	5,512	241	5,872	361	0.14	12.87	361	50.30	-121.467
51	Salmon River at Somes Bar	4,261	184	8,356	751	8,856	500	0.00	1.20	511	45.34	-123.190
52	Santa Cruz Creek near Santa Ynez	3,355	57	5,774	74	6,567	794	0.02	0.00	794	101.41	-119.787
53	Salsipuedes Creek near Lompoc	920	48	1,565	47	1,807	242	0.00	0.00	242	32.51	-120.359
54	Trinity River above Coffee Creek near Trinity Center	5,340	88	6,457	148	9,020	2,563	0.11	32.56	2,573	73.14	-122.640
55	Scott River near Fort Jones	4,333	170	5,850	662	8,482	2,631	0.18	11.57	2,631	34.39	-122.837

Appendix 2-1. Basin characteristics for all 50 sites included in this study of the Central Valley region of California.—Continued

[Site locations are shown in figure 1. See table 2 for definition of basin characteristics. Abbreviations: F, Fahrenheit; –, could not be determined]

Site number	Site name	CENTROIDY (decimal degrees)	OUTLETX (meters)	OUTLETY (meters)	DIST2COAST (miles)	BSLDEM30M (percent)	FOREST (percent)	IMPNLCD01 (percent)	PRECIP (inches)	JANMAXTMP (degrees F)	JANMINTMP (degrees F)	NL Elev (feet)
1	Sacramento River at Shasta Dam	41.140	-2,180,430	2,233,140	80.67	27.75	19.11	0.17	38.15	50.14	33.03	1.000
3	Cottonwood Creek near Cottonwood	40.311	-2,176,350	2,245,860	109.59	16.93	37.31	0.28	47.43	51.08	33.47	0.003
4	Cow Creek near Millville	40.658	-2,172,660	2,232,360	124.56	16.31	45.74	0.28	47.01	46.58	28.77	0.004
5	Battle Creek below Coleman Fish Hatchery	40.446	-2,173,050	2,192,520	133.10	32.67	36.73	0.14	56.86	46.85	29.18	0.988
6	Mill Creek near Los Molinos	40.239	-2,213,730	2,200,500	94.37	34.21	23.24	0.11	36.89	48.59	33.81	0.957
7	Elder Creek near Paskenta	39.996	-2,219,220	2,186,130	93.12	35.64	33.37	0.14	40.31	46.21	31.49	0.105
8	Thomes Creek at Paskenta	39.915	-2,168,160	2,186,400	138.69	28.16	53.98	0.23	58.74	46.04	27.81	0.996
9	Deer Creek near Vina	40.177	-2,159,430	2,156,220	133.18	30.36	69.7	0.16	64.06	49.63	32.77	0.998
10	Big Chico Creek near Chico	39.996	-2,206,050	2,174,400	80.87	28.63	20.38	0.15	31	51.51	34.31	0.159
11	Stony Creek at Black Butte Dam	39.567	-2,157,240	2,149,860	134.31	25.5	69.45	0.52	69.31	48.28	30.94	0.008
12	Butte Creek near Chico	39.965	-2,120,670	2,101,050	144.90	34.27	67.23	0.15	70.38	45.88	27.02	0.770
13	Feather River At Oroville Dam	39.954	–	–	–	–	–	–	–	–	–	1.000
14	North Yuba River at Bullards Bar Dam	39.585	-2,154,060	2,064,600	120.14	19.85	56.54	1.26	47.19	52.97	33.24	1.000
15	Bear River near Wheatland	39.136	-2,244,930	2,099,370	56.08	34.37	12.09	0.08	42.15	52.02	35.71	0.004
16	North Fork Cache Creek at Indian Valley Dam	39.176	-2,141,970	2,027,220	130.47	28.6	57.85	0.68	53.16	47.22	28.61	0.023
17	American River at Fair Oaks	38.941	-2,041,350	1,783,620	153.51	41.24	32.93	0.04	37.96	42.08	20.27	1.000
18	Kings River at Pine Flat Dam	36.919	-2,068,320	1,810,110	145.93	30.97	33.81	0.09	40.29	44.02	22.17	1.000
19	San Joaquin River at Friant Dam	37.337	–	–	–	–	–	–	–	–	–	1.000
20	Chowchilla River at Buchanan Dam	37.386	–	–	–	–	–	–	–	–	–	0.002
23	Del Puerto Creek near Patterson	37.442	-2,182,170	1,896,420	44.47	36.35	9.84	0.07	16.89	54.08	38.15	0.000
24	Merced River at Exchequer Dam	37.670	-2,100,780	1,885,770	125.56	33.78	42.32	0.13	41.22	48.47	25.86	1.000
25	Tuolumne River at New Don Pedro Dam	37.959	-2,109,420	1,901,430	129.76	30.46	37.04	0.32	43.33	45.99	23.22	1.000
26	Stanislaus River at New Melones Dam	38.257	-2,110,920	1,931,040	133.69	29.75	47.81	0.39	45.62	45.02	24.36	1.000
28	Duck Creek near Farmington	37.972	-2,147,370	1,939,470	88.14	3.85	0	0.11	15.12	53.55	36.34	0.000
30	Calaveras River at New Hogan Dam	38.197	-2,130,570	1,959,330	108.03	24.02	47.99	0.58	34.01	53.58	34.45	0.001
31	Mokelumne River at Camanche Dam	38.447	-2,146,290	1,972,020	126.80	28.05	48.99	0.3	46.44	46.51	27.21	1.000
32	Cosumnes River at Michigan Bar	38.603	-2,138,670	2,002,350	115.87	22.46	57.67	0.46	42.39	50.76	31.51	0.134
33	Fresno River near Knowles	37.359	-2,067,630	1,837,350	121.49	22.73	52.23	0.92	33.72	54.82	32.59	0.217
34	South Yuba River at Jones Bar	39.365	-2,120,430	2,089,380	144.48	27.43	55.76	0.41	67.77	44.07	24.24	1.000
35	Middle Yuba River below Our House Dam	39.468	-2,108,190	2,099,940	146.74	32.97	63.15	0.15	68.58	44.32	24.86	1.000
36	Kaweah River at Terminus Dam	36.496	-2,024,040	1,731,690	138.35	41.66	45.98	0.09	37.13	46.82	26.69	1.000
37	Tule River at Success Dam	36.137	-2,026,500	1,691,190	126.86	35.92	48.81	0.2	30.31	50.43	32.15	0.963
38	Kern River at Isabella Dam	36.038	-1,999,050	1,636,440	143.89	33.21	19.99	0.14	22.43	44.49	33.67	1.000
39	Mill Creek near Piedra	36.758	-2,042,130	1,782,480	133.02	30.06	36.38	0.23	28.41	55.36	35.27	0.024
40	Dry Creek near Lemoncove	36.573	-2,025,120	1,735,380	131.21	32.31	35.38	0.1	29.77	55.35	37.42	0.027
41	Deer Creek near Fountain Springs	35.913	-2,020,890	1,676,010	121.70	37.71	54.92	0.06	28.7	51.04	31.82	0.967
42	White River near Ducor	35.824	-2,032,740	1,663,770	114.54	26.98	21.08	0.07	20.61	55.31	35.86	0.009
43	Cache Creek at Clear Lake	39.039	-2,252,190	2,083,290	42.60	21.41	19.09	0.92	40.76	53.12	34.82	0.001
44	Putah Creek at Monticello Dam	38.688	-2,226,030	2,028,090	42.07	25.42	16.45	0.3	34.76	54.44	36.35	0.000
45	Middle Fork Eel River near Dos Rios	39.789	-2,289,960	2,185,590	38.64	32.36	35.28	0.08	56.41	48.6	31.81	0.734
46	South Fork Eel River near Miranda	39.900	-2,311,290	2,247,600	9.06	35.41	52.4	0.17	75.91	53.14	35.66	0.000
47	Mad River above Ruth Reservoir near Forest Glen	40.193	-2,272,380	2,247,930	40.58	36.5	56.25	0.05	65.15	47.59	29.18	0.756
48	East Fork Russian River near Calpella	39.286	-2,288,550	2,131,440	34.85	27.61	19.22	0.22	44.94	53.9	34.74	0.000
49	Salinas River near Pozo	35.293	-2,176,260	1,641,210	22.11	27.97	16.6	0.03	27.85	57.4	36.95	0.003
50	Arroyo Seco near Soledad	36.232	-2,227,350	1,768,680	11.92	45.68	28.82	0.05	31.57	56.86	36.77	0.012
51	Salmon River at Somes Bar	41.293	-2,248,980	2,368,620	45.68	51.94	61.42	0.03	62.06	42.26	28.29	0.999
52	Santa Cruz Creek near Santa Ynez	34.657	-2,152,590	1,553,880	16.60	42.82	30.76	0.02	32.66	56.58	35.18	0.349
53	Salsipuedes Creek near Lompoc	34.553	-2,196,900	1,564,530	6.44	24.94	22.71	0.18	21.04	63.11	43.35	0.000
54	Trinity River above Coffee Creek near Trinity Center	41.229	-2,195,790	2,322,030	74.41	38.74	52.72	0.24	58.25	38.86	26.41	1.000
55	Scott River near Fort Jones	41.480	-2,203,890	2,386,080	63.22	33.16	30.04	0.24	33.18	41.31	26.55	1.000

Appendix 2-2. Summary of censoring decisions for each site and duration in the Central Valley region of California.

[**Abbreviations**: Cens, censored; EMA, Expected Moments Algorithm]

Site number	Site name	Number of years of record	Type of Censoring	Number of censored flows for the indicated duration				
				1-Day	3-Day	7-Day	15-Day	30-Day
1	Sacramento River at Shasta Dam	[1] 77	EMA Cens/ zeros	1	0	0	0	0
			Additional censored	2	2	2	2	2
			Total	3	2	2	2	2
3	Cottonwood Creek near Cottonwood	68	EMA Cens/ zeros	1	1	1	1	1
			Additional censored	0	0	0	0	0
			Total	1	1	1	1	1
4	Cow Creek near Millville	59	EMA Cens/ zeros	1	1	1	1	1
			Additional censored	0	0	0	0	0
			Total	1	1	1	1	1
5	Battle Creek below Coleman Fish Hatchery	68	EMA Cens/ zeros	1	0	0	0	0
			Additional censored	0	1	1	1	1
			Total	1	1	1	1	1
6	Mill Creek near Los Molinos	80	EMA Cens/ zeros	1	1	1	1	1
			Additional censored	0	0	0	0	0
			Total	1	1	1	1	1
7	Elder Creek near Paskenta	60	EMA Cens/ zeros	0	0	1	1	1
			Additional censored	1	1	0	0	0
			Total	1	1	1	1	1
8	Thomes Creek at Paskenta	76	EMA Cens/ zeros	1	1	1	1	1
			Additional censored	0	0	0	0	0
			Total	1	1	1	1	1
9	Deer Creek near Vina	92	EMA Cens/ zeros	1	1	1	1	1
			Additional censored	0	0	0	0	0
			Total	1	1	1	1	1
10	Big Chico Creek near Chico	77	EMA Cens/ zeros	1	1	1	1	1
			Additional censored	0	0	0	0	0
			Total	1	1	1	1	1
11	Stony Creek at Black Butte Dam	[1] 108	EMA Cens/ zeros	1	1	1	1	1
			Additional censored	0	0	0	0	0
			Total	1	1	1	1	1
12	Butte Creek near Chico	78	EMA Cens/ zeros	1	1	1	1	1
			Additional censored	0	0	0	0	0
			Total	1	1	1	1	1
13	Feather River at Oroville Dam	107	EMA Cens/ zeros	0	0	0	0	0
			Additional censored	0	0	0	0	0
			Total	0	0	0	0	0
14	North Yuba at Bullards Bar Dam	68	EMA Cens/ zeros	0	0	0	1	1
			Additional censored	2	2	2	1	1
			Total	2	2	2	2	2
15	Bear River near Wheatland	103	EMA Cens/ zeros	0	0	1	1	1
			Additional censored	1	1	0	0	0
			Total	1	1	1	1	1
16	North Fork Cache Creek at Indian Valley Dam	[1] 78	EMA Cens/ zeros	1	1	1	1	1
			Additional censored	3	3	3	3	3
			Total	4	4	4	4	4

Appendix 2-2.   Summary of censoring decisions at each site and duration in the Central Valley region of California.—Continued

[**Abbreviations**: Cens, censored; EMA, Expected Moments Algorithm]

Site number	Site name	Number of years of record	Type of Censoring	Number of censored flows for the indicated duration				
				1-Day	3-Day	7-Day	15-Day	30-Day
17	American River at Fair Oaks	104	EMA Cens/ zeros	0	0	0	1	1
			Additional censored	1	1	1	0	0
			Total	1	1	1	1	1
18	Kings River at Pine Flat Dam	113	EMA Cens/ zeros	0	0	0	0	0
			Additional censored	0	0	0	0	0
			Total	0	0	0	0	0
19	San Joaquin River at Friant Dam	105	EMA Cens/ zeros	0	0	0	0	0
			Additional censored	0	0	0	0	0
			Total	0	0	0	0	0
20	Chowchilla River at Buchanan Dam	[1] 80	EMA Cens/ zeros	1	0	0	0	0
			Additional censored	0	1	1	1	1
			Total	1	1	1	1	1
23	Del Puerto Creek near Patterson	44	EMA Cens/ zeros	1	0	0	0	1
			Additional censored	0	1	1	1	0
			Total	1	1	1	1	1
24	Merced River at Exchequer Dam	[1] 107	EMA Cens/ zeros	1	1	1	1	1
			Additional censored	0	0	0	0	0
			Total	1	1	1	1	1
25	Tuolumne River at New Don Pedro Dam	112	EMA Cens/ zeros	0	0	0	0	0
			Additional censored	0	0	0	0	0
			Total	0	0	0	0	0
26	Stanislaus River at New Melones Dam	[1] 93	EMA Cens/ zeros	0	0	0	0	0
			Additional censored	1	1	1	1	1
			Total	1	1	1	1	1
28	Duck Creek near Farmington	30	EMA Cens/ zeros	1	1	0	0	0
			Additional censored	0	0	1	1	1
			Total	1	1	1	1	1
30	Calaveras River at New Hogan Dam	96	EMA Cens/ zeros	1	1	1	1	1
			Additional censored	0	0	0	0	0
			Total	1	1	1	1	1
31	Mokelumne River at Camanche Dam	[1] 104	EMA Cens/ zeros	0	0	0	0	0
			Additional censored	0	0	0	0	0
			Total	0	0	0	0	0
32	Cosumnes River at Michigan Bar	101	EMA Cens/ zeros	1	1	1	1	1
			Additional censored	0	0	0	0	0
			Total	1	1	1	1	1
33	Fresno River near Knowles	76	EMA Cens/ zeros	0	0	0	0	0
			Additional censored	0	0	0	0	0
			Total	0	0	0	0	0
34	South Yuba River at Jones Bar	57	EMA Cens/ zeros	0	0	0	1	1
			Additional censored	1	1	1	0	0
			Total	1	1	1	1	1
35	Middle Yuba River below Our House Dam	37	EMA Cens/ zeros	0	0	0	0	0
			Additional censored	0	0	0	0	0
			Total	0	0	0	0	0

Appendix 2-2.   Summary of censoring decisions at each site and duration in the Central Valley region of California.—Continued

[**Abbreviations**: Cens, censored; EMA, Expected Moments Algorithm]

Site number	Site name	Number of years of record	Type of Censoring	Number of censored flows for the indicated duration				
				1-Day	3-Day	7-Day	15-Day	30-Day
36	Kaweah River at Terminus Dam	50	EMA Cens/ zeros	0	0	0	0	0
			Additional censored	0	0	0	0	0
			Total	0	0	0	0	0
37	Tule River at Success Dam	50	EMA Cens/ zeros	0	0	0	0	0
			Additional censored	0	0	0	0	0
			Total	0	0	0	0	0
38	Kern River at Isabella Dam	114	EMA Cens/ zeros	0	0	0	0	0
			Additional censored	0	0	0	0	0
			Total	0	0	0	0	0
39	Mill Creek near Piedra	52	EMA Cens/ zeros	0	0	0	0	0
			Additional censored	0	0	0	0	0
			Total	0	0	0	0	0
40	Dry Creek near Lemoncove	50	EMA Cens/ zeros	0	0	1	1	0
			Additional censored	1	1	0	0	1
			Total	1	1	1	1	1
41	Deer Creek near Fountain Springs	41	EMA Cens/ zeros	0	0	0	0	0
			Additional censored	0	0	0	0	0
			Total	0	0	0	0	0
42	White River near Ducor	46	EMA Cens/ zeros	0	0	0	0	0
			Additional censored	0	0	0	0	0
			Total	0	0	0	0	0
43	Cache Creek at Clear Lake	87	EMA Cens/ zeros	1	1	1	1	1
			Additional censored	3	3	3	3	3
			Total	4	4	4	4	4
44	Putah Creek at Monticello Dam	78	EMA Cens/ zeros	1	1	2	2	2
			Additional censored	11	11	9	9	9
			Total	12	12	11	11	11
45	Middle Fork Eel River near Dos Rios	43	EMA Cens/ zeros	1	1	1	1	1
			Additional censored	0	0	0	0	0
			Total	1	1	1	1	1
46	South Fork Eel River near Miranda	68	EMA Cens/ zeros	1	1	1	1	1
			Additional censored	0	0	0	0	0
			Total	1	1	1	1	1
47	Mad River above Ruth Reservoir near Forest Glen	28	EMA Cens/ zeros	0	0	0	0	0
			Additional censored	0	0	0	0	0
			Total	0	0	0	0	0
48	East Fork Russian River near Calpella	67	EMA Cens/ zeros	1	1	1	1	1
			Additional censored	0	0	0	0	0
			Total	1	1	1	1	1
49	Salinas River near Pozo	41	EMA Cens/ zeros	0	0	0	0	0
			Additional censored	0	0	0	0	0
			Total	0	0	0	0	0
50	Arroyo Seco near Soledad	[1] 107	EMA Cens/ zeros	0	1	1	1	1
			Additional censored	1	0	0	0	0
			Total	1	1	1	1	1

Appendix 2-2. Summary of censoring decisions at each site and duration in the Central Valley region of California.—Continued

[**Abbreviations:** Cens, censored; EMA, Expected Moments Algorithm]

Site number	Site name	Number of years of record	Type of Censoring	Number of censored flows for the indicated duration				
				1-Day	3-Day	7-Day	15-Day	30-Day
51	Salmon River at Somes Bar	84	EMA Cens/ zeros	1	1	1	1	1
			Additional censored	0	0	0	0	0
			Total	1	1	1	1	1
52	Santa Cruz Creek near Santa Ynez	67	EMA Cens/ zeros	2	2	1	0	0
			Additional censored	4	3	4	5	4
			Total	6	5	5	5	4
53	Salsipuedes Creek near Lompoc	67	EMA Cens/ zeros	0	0	0	0	0
			Additional censored	0	0	0	0	0
			Total	0	0	0	0	0
54	Trinity River above Coffee Creak near Trinity Center	51	EMA Cens/ zeros	0	1	1	1	1
			Additional censored	1	0	0	0	0
			Total	1	1	1	1	1
55	Scott River near Fort Jones	67	EMA Cens/ zeros	1	1	1	1	1
			Additional censored	0	0	0	0	0
			Total	1	1	1	1	1

[1] The period of record and number of years of record could be less than the given value for selected durations.

Appendix 2-3.   Sample skew for each site and duration used in the regional-skew analyses for the Central Valley region of California.

Site number	Site name	Number of years of record	Sample log-space skew for indicated duration				
			1-Day	3-Day	7-Day	15-Day	30-Day
1	Sacramento River at Shasta Dam	77	−0.104	−0.467	−0.282	−0.163	−0.368
3	Cottonwood Creek near Cottonwood	68	−0.556	−0.579	−0.417	−0.568	−0.594
4	Cow Creek near Millville	59	−0.762	−0.623	−0.400	−0.316	−0.350
5	Battle Creek below Coleman Fish Hatchery	68	−0.284	−0.065	0.039	0.017	0.046
6	Mill Creek near Los Molinos	80	0.009	−0.029	0.014	−0.055	−0.052
7	Elder Creek near Paskenta	60	−1.092	−1.007	−0.731	−0.881	−0.972
8	Thomes Creek at Paskenta	76	−0.069	−0.035	−0.069	−0.197	−0.411
9	Deer Creek near Vina	92	−0.286	−0.223	−0.189	−0.198	−0.247
10	Big Chico Creek near Chico	77	−0.936	−0.700	−0.454	−0.523	−0.584
11	Stony Creek at Black Butte Dam	[1] 108	−0.714	−0.796	−0.407	−0.606	−0.762
12	Butte Creek near Chico	78	−0.241	−0.246	−0.140	−0.157	−0.167
13	Feather River at Oroville Dam	107	−0.240	−0.206	−0.224	−0.332	−0.412
14	North Yuba River at Bullards Bar Dam	68	−0.030	0.107	0.088	−0.154	−0.315
15	Bear River near Wheatland	103	−0.780	−0.747	−0.627	−0.750	−0.839
16	North Fork Cache Creek at Indian Valley Dam	[1] 78	−0.841	−1.008	−0.824	−0.810	−0.830
17	American River at Fair Oaks	104	−0.066	0.021	−0.001	−0.131	−0.308
18	Kings River at Pine Flat Dam	113	0.148	0.227	0.205	0.190	0.100
19	San Joaquin River at Friant Dam	105	0.186	0.222	0.182	−0.002	−0.058
20	Chowchilla River at Buchanan Dam	[1] 80	−0.843	−0.787	−0.689	−0.621	−0.534
23	Del Puerto Creek near Patterson	44	−0.852	−0.974	−0.806	−0.809	−0.754
24	Merced River at Exchequer Dam	[1] 107	−0.228	−0.104	−0.085	−0.238	−0.343
25	Tuolumne River at New Don Pedro Dam	112	−0.183	−0.120	−0.160	−0.349	−0.480
26	Stanislaus River at New Melones Dam	[1] 93	0.175	0.192	0.180	0.010	−0.037
28	Duck Creek near Farmington	30	−0.744	−0.823	−1.003	−1.032	−1.029
30	Calaveras River at New Hogan Dam	96	−1.048	−0.870	−0.855	−0.728	−0.725
31	Mokelumne River at Camanche Dam	[1] 104	0.088	0.082	0.035	−0.144	−0.301
32	Cosumnes River at Michigan Bar	101	−0.588	−0.533	−0.514	−0.558	−0.603
33	Fresno River near Knowles	76	−0.187	−0.230	−0.303	−0.315	−0.284
34	South Yuba River at Jones Bar	57	−0.005	0.198	0.182	0.055	0.048
35	Middle Yuba River below Our House Dam	37	−0.319	−0.188	−0.024	0.110	0.172
36	Kaweah River at Terminus Dam	50	0.194	0.217	0.189	0.111	0.000
37	Tule River at Success Dam	50	0.057	0.027	0.020	−0.001	0.036
38	Kern River at Isabella Dam	114	0.282	0.221	0.180	0.145	0.080
39	Mill Creek near Piedra	52	−0.192	−0.256	−0.198	−0.141	−0.113
40	Dry Creek near Lemoncove	50	−0.534	−0.594	−0.504	−0.391	−0.427
41	Deer Creek near Fountain Springs	41	−0.046	0.013	0.095	0.179	0.248
42	White River near Ducor	46	−0.317	−0.181	−0.088	0.004	0.089
43	Cache Creek at Clear Lake	87	−0.466	−0.442	−0.398	−0.813	−0.799
44	Putah Creek at Monticello Dam	78	−1.010	−0.741	−0.573	−0.798	−0.455
45	Middle Fork Eel River near Dos Rios	43	−0.312	−0.262	−0.292	−0.442	−0.680
46	South Fork Eel River near Miranda	68	−0.096	−0.050	−0.220	−0.423	−0.449
47	Mad River above Ruth Reservoir near Forest Glen	28	−0.473	−0.182	0.011	0.028	−0.230
48	East Fork Russian River near Calpella	67	−0.330	−0.179	−0.134	−0.441	−0.480
49	Salinas River near Pozo	41	−0.730	−0.611	−0.528	−0.478	−0.371
50	Arroyo Seco near Soledad	[1] 107	−0.608	−0.717	−0.743	−0.738	−0.651
51	Salmon River at Somes Bar	84	−0.073	−0.032	0.041	−0.091	−0.259
52	Santa Cruz Creek near Santa Ynez	67	−0.526	−0.483	−0.417	−0.345	−0.402
53	Salsipuedes Creek near Lompoc	67	−0.608	−0.472	−0.361	−0.297	−0.193
54	Trinity River above Coffee Creek near Trinity Center	51	−0.069	−0.061	−0.043	−0.258	−0.276
55	Scott River near Fort Jones	67	−0.277	−0.246	−0.278	−0.321	−0.354

[1] The period of record and number of years of record could be less than the given value for selected durations.

Appendix 2-4. Regional skew for each site and duration as determined by the regional-skew analyses for the Central Valley region of California.

Site number	Site name	Regional log-space skew for the indicated duration				
		1-Day	3-Day	7-Day	15-Day	30-Day
1	Sacramento River at Shasta Dam	−0.049	−0.008	0.002	−0.096	−0.192
3	Cottonwood Creek near Cottonwood	−0.733	−0.688	−0.586	−0.644	−0.632
4	Cow Creek near Millville	−0.732	−0.688	−0.586	−0.643	−0.632
5	Battle Creek below Coleman Fish Hatchery	−0.057	−0.017	−0.005	−0.103	−0.197
6	Mill Creek near Los Molinos	−0.078	−0.037	−0.023	−0.119	−0.211
7	Elder Creek near Paskenta	−0.662	−0.619	−0.526	−0.587	−0.587
8	Thomes Creek at Paskenta	−0.052	−0.011	0.000	−0.098	−0.194
9	Deer Creek near Vina	−0.050	−0.010	0.001	−0.097	−0.193
10	Big Chico Creek near Chico	−0.626	−0.582	−0.494	−0.558	−0.563
11	Stony Creek at Black Butte Dam	−0.729	−0.685	−0.583	−0.641	−0.629
12	Butte Creek near Chico	−0.207	−0.166	−0.134	−0.223	−0.294
13	Feather River at Oroville Dam	−0.049	−0.008	0.002	−0.096	−0.192
14	North Yuba River at Bullards Bar Dam	−0.049	−0.008	0.002	−0.096	−0.192
15	Bear River near Wheatland	−0.732	−0.688	−0.586	−0.643	−0.632
16	North Fork Cache Creek at Indian Valley Dam	−0.719	−0.675	−0.574	−0.633	−0.623
17	American River at Fair Oaks	−0.049	−0.008	0.002	−0.096	−0.192
18	Kings River at Pine Flat Dam	−0.049	−0.008	0.002	−0.096	−0.192
19	San Joaquin River at Friant Dam	−0.049	−0.008	0.002	−0.096	−0.192
20	Chowchilla River at Buchanan Dam	−0.733	−0.689	−0.586	−0.644	−0.632
23	Del Puerto Creek near Patterson	−0.734	−0.690	−0.588	−0.645	−0.633
24	Merced River at Exchequer Dam	−0.049	−0.008	0.002	−0.096	−0.192
25	Tuolumne River at New Don Pedro Dam	−0.049	−0.008	0.002	−0.096	−0.192
26	Stanislaus River at New Melones Dam	−0.049	−0.008	0.002	−0.096	−0.192
28	Duck Creek near Farmington	−0.735	−0.691	−0.588	−0.645	−0.633
30	Calaveras River at New Hogan Dam	−0.734	−0.690	−0.587	−0.645	−0.633
31	Mokelumne River at Camanche Dam	−0.049	−0.008	0.002	−0.096	−0.192
32	Cosumnes River at Michigan Bar	−0.642	−0.599	−0.508	−0.571	−0.574
33	Fresno River near Knowles	−0.586	−0.543	−0.460	−0.526	−0.538
34	South Yuba River at Jones Bar	−0.049	−0.008	0.002	−0.096	−0.192
35	Middle Yuba River below Our House Dam	−0.049	−0.008	0.002	−0.096	−0.192
36	Kaweah River at Terminus Dam	−0.049	−0.008	0.002	−0.096	−0.192
37	Tule River at Success Dam	−0.074	−0.034	−0.020	−0.117	−0.209
38	Kern River at Isabella Dam	−0.049	−0.008	0.002	−0.096	−0.192
39	Mill Creek near Piedra	−0.718	−0.674	−0.574	−0.632	−0.623
40	Dry Creek near Lemoncove	−0.716	−0.672	−0.572	−0.630	−0.621
41	Deer Creek near Fountain Springs	−0.071	−0.031	−0.017	−0.114	−0.206
42	White River near Ducor	−0.728	−0.684	−0.582	−0.640	−0.629
43	Cache Creek at Clear Lake	−0.734	−0.690	−0.587	−0.645	−0.633
44	Putah Creek at Monticello Dam	−0.735	−0.691	−0.588	−0.645	−0.633
45	Middle Fork Eel River near Dos Rios	−0.231	−0.190	−0.155	−0.242	−0.309
46	South Fork Eel River near Miranda	−0.735	−0.690	−0.588	−0.645	−0.633
47	Mad River above Ruth Reservoir near Forest Glen	−0.216	−0.175	−0.142	−0.230	−0.300
48	East Fork Russian River near Calpella	−0.735	−0.690	−0.588	−0.645	−0.633
49	Salinas River near Pozo	−0.733	−0.689	−0.586	−0.644	−0.632
50	Arroyo Seco near Soledad	−0.726	−0.682	−0.581	−0.639	−0.628
51	Salmon River at Somes Bar	−0.049	−0.009	0.002	−0.096	−0.192
52	Santa Cruz Creek near Santa Ynez	−0.495	−0.453	−0.382	−0.454	−0.479
53	Salsipuedes Creek near Lompoc	−0.735	−0.691	−0.588	−0.645	−0.633
54	Trinity River above Coffee Creek near Trinity Center	−0.049	−0.008	0.002	−0.096	−0.192
55	Scott River near Fort Jones	−0.049	−0.008	0.002	−0.096	−0.192

Appendix 2-5. Variance of prediction (VP_{old}) for each site and duration as determined by the regional-skew analyses for the Central Valley region of California.

Site number	Site name	VP_{old} for indicated duration				
		1-Day	3-Day	7-Day	15-Day	30-Day
1	Sacramento River at Shasta Dam	0.0384	0.0396	0.0401	0.0404	0.0468
3	Cottonwood Creek near Cottonwood	0.0572	0.0587	0.0578	0.0614	0.0654
4	Cow Creek near Millville	0.0573	0.0588	0.0578	0.0615	0.0655
5	Battle Creek below Coleman Fish Hatchery	0.0386	0.0398	0.0403	0.0406	0.0470
6	Mill Creek near Los Molinos	0.0387	0.0399	0.0404	0.0409	0.0472
7	Elder Creek near Paskenta	0.0543	0.0557	0.0551	0.0583	0.0627
8	Thomes Creek at Paskenta	0.0384	0.0397	0.0401	0.0405	0.0468
9	Deer Creek near Vina	0.0382	0.0395	0.0400	0.0404	0.0467
10	Big Chico Creek near Chico	0.0526	0.0541	0.0536	0.0566	0.0611
11	Stony Creek at Black Butte Dam	0.0571	0.0585	0.0578	0.0614	0.0653
12	Butte Creek near Chico	0.0406	0.0420	0.0424	0.0433	0.0494
13	Feather River at Oroville Dam	0.0380	0.0393	0.0399	0.0403	0.0465
14	North Yuba River at Bullards Bar Dam	0.0385	0.0397	0.0402	0.0405	0.0469
15	Bear River near Wheatland	0.0568	0.0583	0.0575	0.0612	0.0650
16	North Fork Cache Creek at Indian Valley Dam	0.0565	0.0580	0.0572	0.0607	0.0647
17	American River at Fair Oaks	0.0381	0.0394	0.0399	0.0403	0.0465
18	Kings River at Pine Flat Dam	0.0380	0.0393	0.0399	0.0402	0.0464
19	San Joaquin River at Friant Dam	0.0381	0.0394	0.0399	0.0403	0.0465
20	Chowchilla River at Buchanan Dam	0.0571	0.0586	0.0577	0.0614	0.0653
23	Del Puerto Creek near Patterson	0.0576	0.0590	0.0581	0.0617	0.0658
24	Merced River at Exchequer Dam	0.0380	0.0393	0.0399	0.0403	0.0465
25	Tuolumne River at New Don Pedro Dam	0.0380	0.0393	0.0399	0.0402	0.0465
26	Stanislaus River at New Melones Dam	0.0382	0.0395	0.0400	0.0403	0.0466
28	Duck Creek near Farmington	0.0578	0.0592	0.0582	0.0618	0.0660
30	Calaveras River at New Hogan Dam	0.0570	0.0585	0.0576	0.0613	0.0651
31	Mokelumne River at Camanche Dam	0.0381	0.0394	0.0399	0.0403	0.0466
32	Cosumnes River at Michigan Bar	0.0530	0.0546	0.0540	0.0572	0.0615
33	Fresno River near Knowles	0.0511	0.0526	0.0522	0.0550	0.0597
34	South Yuba River at Jones Bar	0.0386	0.0398	0.0403	0.0405	0.0470
35	Middle Yuba River below Our House Dam	0.0388	0.0400	0.0404	0.0406	0.0472
36	Kaweah River at Terminus Dam	0.0387	0.0399	0.0403	0.0406	0.0471
37	Tule River at Success Dam	0.0389	0.0402	0.0406	0.0410	0.0474
38	Kern River at Isabella Dam	0.0380	0.0393	0.0399	0.0402	0.0464
39	Mill Creek near Piedra	0.0568	0.0582	0.0573	0.0609	0.0650
40	Dry Creek near Lemoncove	0.0567	0.0581	0.0573	0.0608	0.0649
41	Deer Creek near Fountain Springs	0.0390	0.0402	0.0406	0.0409	0.0475
42	White River near Ducor	0.0573	0.0587	0.0578	0.0614	0.0655
43	Cache Creek at Clear Lake	0.0571	0.0585	0.0577	0.0614	0.0652
44	Putah Creek at Monticello Dam	0.0575	0.0589	0.0580	0.0616	0.0656
45	Middle Fork Eel River near Dos Rios	0.0413	0.0427	0.0430	0.0440	0.0501
46	South Fork Eel River near Miranda	0.0573	0.0588	0.0579	0.0615	0.0655
47	Mad River above Ruth Reservoir near Forest Glen	0.0412	0.0425	0.0428	0.0437	0.0499
48	East Fork Russian River near Calpella	0.0573	0.0588	0.0579	0.0615	0.0655
49	Salinas River near Pozo	0.0576	0.0590	0.0580	0.0616	0.0657
50	Arroyo Seco near Soledad	0.0565	0.0580	0.0572	0.0609	0.0647
51	Salmon River at Somes Bar	0.0383	0.0395	0.0401	0.0404	0.0467
52	Santa Cruz Creek near Santa Ynez	0.0481	0.0496	0.0494	0.0517	0.0568
53	Salsipuedes Creek near Lompoc	0.0573	0.0588	0.0579	0.0615	0.0655
54	Trinity River above Coffee Creak near Trinity Center	0.0387	0.0399	0.0403	0.0406	0.0471
55	Scott River near Fort Jones	0.0385	0.0397	0.0402	0.0405	0.0469

Appendix 3. Methodology for Regional-Skew Analysis for Rainfall Floods of Differing Durations

Parrett and others (2011) showed in the California Annual Peak Flow Study that the cross correlations among annual peak discharges in California are often greater than those reported in other studies. This presents difficulties in the regional-skew analysis because a Bayesian generalized least squares (B-GLS) analysis seeks to exploit the cross correlations among the sample skews to obtain the best possible estimates of the model parameters. If the cross correlations are high, the generalized least square (GLS) estimators can become relatively complicated as a result of the effort to find the most efficient estimator of the parameters. Unfortunately, the precision of the estimated cross correlation between any two sites is not sufficient to justify the sophisticated weights (both positive and negative) that the B-GLS analysis generates. Thus, an alternate model fitting procedure using both weighted least squares (WLS) and generalized least squares (GLS) was developed so that the regional-skew analysis would provide stable and defensible results.

In this study, which considered floods of different durations, the cross correlations among flows were even greater. An alternative procedure was developed to identify a regional-skew model for each flood duration. This alternative procedure uses an ordinary least squares (OLS) analysis to generate an initial regional-skew model, which is used to generate a regional-skew estimate for each site. That OLS skew estimate is used to compute the sampling variance of each skew estimator for use in a WLS analysis. Then, WLS is used to generate the estimator of the regional-skew model parameters. Finally, B-GLS is used to estimate the precision of that parameter estimator and to estimate the model error variance. The three-step procedure was repeated to develop a regional-skew model and the associated error analysis for each flood duration.

Step 1: Ordinary Least Squares Analysis

The first step in the regional-skew analysis is the estimation of a regional-skew model by using OLS. This is an iterative procedure. For the first iteration, the constant model is used. After the subsequent WLS and GLS analyses determine which basin characteristics are statistically significant in explaining regional skew, the OLS regional model can be expanded to incorporate those additional basin characteristics. The OLS analysis estimates the regional regression parameters, which can then be used to generate stable regional-skew estimates for each site in the study. The OLS analysis uses unbiased at-site sample-skew estimators.

The at-site skews are unbiased by using the correction factor developed by Tasker and Stedinger (1986) and employed by Reis and others (2005). The unbiased at-site skew estimator is as follows:

$$\hat{\gamma}_i = \left[1 + \frac{6}{N_i}\right] G_i \tag{3-1}$$

where, $\hat{\gamma}_i$ is the unbiased at-site sample skew for site i, N_i is the systematic record length at site i, and G_i is the traditional biased at-site skew estimator for site i or the expected moments algorithm (EMA) estimate if the site has zero flows, low outliers, or historical peaks. When unbiasing the skew, N_i, is the number of systematic peaks. Thus, additional information provided by any historical flood period is neglected.

The regional regression parameters estimated by OLS, $\hat{\beta}_{OLS}$, are calculated as follows:

$$\hat{\beta}_{OLS} = \left(\mathbf{X}^T\mathbf{X}\right)^{-1}\mathbf{X}^T\hat{\gamma} \tag{3-2}$$

where the superscript T denotes a matrix transpose, $\mathbf{X}$ is the ($n \times k$) matrix of basin characteristics, $\hat{\gamma}$ is the ($n \times 1$) vector of the unbiased at-site sample skews, n is the number of gage sites, and k is the number of basin parameters, including a column of ones to estimate the constant. After computing $\hat{\beta}_{OLS}$, the unbiased regional estimate of the skew for each site is given by the following:

$$\hat{\gamma}_{OLS} = \mathbf{X}\hat{\beta}_{OLS} \tag{3-3}$$

These estimated regional-skew values, $\hat{\gamma}_{OLS}$, are then used with the at-site record lengths to estimate the variance of the at-site sample skew. The variance of the unbiased at-site skew estimators is calculated by the following equation:

$$Var[\hat{\gamma}_i] = \left[1 + \frac{6}{N_i}\right]^2 Var[G_i] \tag{3-4}$$

The variance of the unbiased at-site regional-skew estimator from equation 3–1 is calculated by using the equations developed by Griffis and Stedinger (2009):

$$Var[\hat{\gamma}_i] = \left(1 + \frac{6}{N_i}\right)^2 \left(\frac{6}{N_i + a(N)}\right)\left(1 + \left(\frac{9}{6} + b(N)\right)(\hat{\gamma}_{OLS,i})^2 + \left(\frac{15}{48} + c(N)\right)(\hat{\gamma}_{OLS,i})^4\right) \tag{3-5}$$

where

$$a = \frac{17.75}{N_i^2} + \frac{50.06}{N_i^3},$$

$$b = \frac{3.92}{N_i^{0.3}} - \frac{31.1}{N_i^{0.6}} + \frac{34.86}{N_i^{0.9}}, \text{ and}$$

$$c = -\frac{7.31}{N_i^{0.59}} + \frac{45.9}{N_i^{1.18}} - \frac{86.5}{N_i^{1.77}}$$

The variances of the at-site skews, calculated with equation 3–5, are based on the regional OLS estimator of the skew coefficient instead of the at-site skew estimator. This makes the weights in the subsequent steps for the at-site skew estimates relatively independent. The computation generally neglects complicating factors, such as zero-flow years, censored observations (low outliers), and historical information.

Step Two: Weighted Least Squares Analysis

A weighted least squares (WLS) analysis is used to develop estimators of the regression coefficients for the regional-skew model for each flood duration. The WLS analysis explicitly reflects variations in record length but neglects cross correlations, thereby avoiding the problems encountered with the GLS analysis. After the regression model coefficients are determined by using WLS, the precision of the model and the precision of the regression coefficients are estimated by using an appropriate GLS analysis.

The first step in the WLS analysis is to use Bayesian-WLS (B-WLS) to estimate the model error variance, denoted $\sigma^2_{\delta,B-WLS}$ (Reis and others, 2005). Using a B-WLS approach to estimate the model error variance will avoid the possible pitfall of estimating the model error variance as zero, which can occur when using method-of-moments WLS. Given the model error variance estimator, $\sigma^2_{\delta,B-WLS}$, a WLS analysis is used to generate the weight matrix, $\mathbf{W}$, needed to compute estimates of the regression parameters $\hat{\boldsymbol{\beta}}_{WLS}$. In order to compute $\mathbf{W}$, a diagonal covariance matrix, $\boldsymbol{\Lambda}_{\mathbf{WLS}}\left(\sigma^2_{\delta,B-WLS}\right)$, is created. As specified in equation 3–6, the diagonal elements of the covariance matrix are the sum of the estimated model error variance, $\sigma^2_{\delta,B-WLS}$, and the variance of the unbiased at-site skew estimator, $Var[\hat{\boldsymbol{\gamma}}]$, which depends on the at-site record length and the estimate of the regional skew for each site calculated by OLS, $\hat{\boldsymbol{\gamma}}_{OLS}$. (See equation 3–5 for the calculation of $Var[\hat{\boldsymbol{\gamma}}]$). The off-diagonal elements of $\boldsymbol{\Lambda}_{\mathbf{WLS}}\left(\sigma^2_{\delta,B-WLS}\right)$ are zero because cross correlations among gage sites are not considered in a WLS analysis. Thus, the $(n \times n)$ covariance matrix, $\boldsymbol{\Lambda}_{\mathbf{WLS}}\left(\sigma^2_{\delta,B-WLS}\right)$, is given by the equation:

$$\boldsymbol{\Lambda}_{\mathbf{WLS}}\left(\sigma^2_{\delta,B-WLS}\right) = \sigma^2_{\delta,B-WLS} \ \mathbf{I} + diag\left(Var[\hat{\boldsymbol{\gamma}}]\right) \tag{3-6}$$

where $\mathbf{I}$ is an $(n \times n)$ identity matrix, n is the number of gage sites in the study, and $diag\left(Var[\hat{\boldsymbol{\gamma}}]\right)$ is an $(n \times n)$ matrix containing the variance of the unbiased at-site sample-skew estimators, $Var[\hat{\boldsymbol{\gamma}}]$, on the diagonal and zeros on the off-diagonal. By using that covariance matrix, the WLS weights are calculated as follows:

$$\mathbf{W} = \left[\mathbf{X}^T \boldsymbol{\Lambda}_{\mathbf{WLS}}\left(\sigma^2_{\delta,B-WLS}\right)^{-1} \mathbf{X}\right]^{-1} \mathbf{X}^T \boldsymbol{\Lambda}_{\mathbf{WLS}}\left(\sigma^2_{\delta,B-WLS}\right)^{-1} \tag{3-7}$$

where $\mathbf{W}$ is the $(k \times n)$ matrix of weights, $\mathbf{X}$ is the $(n \times k)$ matrix of basin parameters, and k is the number of columns in the $\mathbf{X}$ matrix. These weights are used to compute the final estimates of the regression parameters $\hat{\boldsymbol{\beta}}$:

$$\hat{\boldsymbol{\beta}}_{WLS} = \mathbf{W}\hat{\boldsymbol{\gamma}} \tag{3-8}$$

where $\hat{\boldsymbol{\beta}}_{WLS}$ is the $(k \times 1)$ vector of estimated regression parameters.

Step Three: Bayesian-Generalized Least Squares Analysis

After the regression model coefficients, $\hat{\boldsymbol{\beta}}_{WLS}$, and weights, $\mathbf{W}$, are determined by using WLS, the precision of the model and the precision of the regression coefficients are estimated by using a B-GLS analysis. Following the B-GLS regression framework for regional skew developed by Reis and others (2005), the posterior probability density function for the model error variance, $\sigma^2_{\delta,B-GLS}$, becomes

$$f\left(\sigma^2_{\delta,B-GLS} \mid \hat{\boldsymbol{\gamma}},\hat{\boldsymbol{\beta}}_{WLS}\right) \propto \xi\left(\sigma^2_{\delta,B-GLS}\right)\left|\boldsymbol{\Lambda}_{\mathbf{GLS}}\left(\sigma^2_{\delta,B-GLS}\right)\right|^{-0.5}$$
$$\exp\left[-0.5\left(\hat{\boldsymbol{\gamma}} - \mathbf{X}\hat{\boldsymbol{\beta}}_{WLS}\right)^T \left(\boldsymbol{\Lambda}_{\mathbf{GLS}}\left(\sigma^2_{\delta,B-GLS}\right)\right)^{-1}\left(\hat{\boldsymbol{\gamma}} - \mathbf{X}\hat{\boldsymbol{\beta}}_{WLS}\right)\right] \tag{3-9}$$

where $\hat{\boldsymbol{\gamma}}$ represents the skew data and $\xi\left(\sigma^2_{\delta,B-GLS}\right)$ is the exponential prior for the model error variance, which has the following form:

$$\xi\left(\sigma^2_{\delta,B-GLS}\right) = \lambda e^{-\lambda\left(\sigma^2_{\delta,B-GLS}\right)}, \ \ \sigma^2_{\delta,B-GLS} > 0 \tag{3-10}$$

A value of 10 was assigned for λ, corresponding to a mean model error variance of one-tenth. The resulting prior assigns a 63-percent probability to the interval, [0, 0.1]; a 86-percent probability to the interval, [0,0.2]; and a 95-percent probability to the interval, [0, 0.3].

The B-GLS model error variance can then be used to compute the precision of the regression parameters, $\hat{\beta}_{WLS}$, that were calculated with the WLS weights, $\mathbf{W}$. The GLS covariance matrix for the WLS β-estimator, $\hat{\beta}_{WLS}$, is simply the following:

$$\Sigma\left(\hat{\beta}_{WLS}\right) = \mathbf{W}\Lambda_{\mathbf{GLS}}\left(\sigma_{\delta,B-GLS}^2\right)\mathbf{W}^T \qquad (3\text{-}11)$$

where $\Lambda_{\mathbf{GLS}}\left(\sigma_{\delta,B-GLS}^2\right)$ is an $(n \times n)$ GLS covariance matrix calculated as follows:

$$\Lambda_{\mathbf{GLS}}\left(\sigma_{\delta,B-GLS}^2\right) = \sigma_{\delta,B-GLS}^2\,\mathbf{I} + \Sigma(\hat{\gamma}) \qquad (3\text{-}12)$$

Here, $\mathbf{I}$ is the $(n \times n)$ identity matrix, and $\Sigma(\hat{\gamma})$ is an $(n \times n)$ matrix containing the sampling variances of the unbiased skewness estimators, $Var[\hat{\gamma}_i]$, and the covariances of the skewness estimators, $\hat{\gamma}_i$. The elements of $\Sigma(\hat{\gamma})$ are determined by the the cross correlation of concurrent systematic annual peak discharges (eq.7) and the cf factor (eq. 8). When calculating the cf factor by using the ratio of the number of concurrent peak flows at a pair of sites to the total number of peak discharges at both sites, only the systematic records are considered. Thus, any additional information provided by a historical flood period included in the EMA analysis is neglected in calculating the cross correlation of peak flows and the cf factor. This was not an issue in this study because no historical flood information was used.

Diagnostic Statistics for WLS/GLS Regional Analysis

This section describes statistics for evaluating the precision of model predictions and whether particular sites have unusual leverage or influence on the results. The variance of prediction is a common metric used to choose the one model among several that provides the most accurate estimator of the dependent variable because it combines both the model error variance and the sampling error in the model parameters.

Variance of Prediction

The variance of prediction depends on whether one is considering a new site, which was not used to derive the estimate of the parameters, or an old site, where the sample estimator of the skew was used to compute the estimates of the parameters. For an old site, there is correlation between error in the at-site estimator and the estimated parameters.

The Bayesian variance of prediction of the skew at a new site with basin characteristics $\mathbf{x}_i$ is given by the equation:

$$VP_{new}(i) = \mathop{E}_{\sigma_{\delta,B-GLS}^2}\left[\sigma_{\delta,B-GLS}^2 + \mathbf{x}_i\mathbf{W}\left(\Lambda_{\mathbf{GLS}}\left(\sigma_{\delta,B-GLS}^2\right)\right)\mathbf{W}^T\mathbf{x}_i^T\right] \qquad (3\text{-}13a)$$

$$VP_{new}(i) = E\left[\sigma_{\delta,B-GLS}^2\right] + \mathbf{x}_i Var\left[\hat{\beta}_{WLS}\right]\mathbf{x}_i^T$$

where $\sigma_{\delta,B-GLS}^2$ reflects the underlying error in the model, and $\mathbf{x}_i\mathbf{W}\Lambda\mathbf{W}^T\mathbf{x}_i^T$ reflects the precision with which the model parameters can be estimated and the possible errors that would occur in predicting the skew at a site with basin characteristics $\mathbf{x}_i$.

However, if the predictions are made for the n old sites that were used in the regression analysis, the Bayesian variance of prediction is given by the equation:

$$VP_{old}(i) = \underset{\sigma^2_{\delta,B-GLS}}{E}\left[\sigma^2_{\delta,B-GLS} + \mathbf{x}_i\mathbf{W}\left(\mathbf{\Lambda_{GLS}}\left(\sigma^2_{\delta,B-GLS}\right)\right)\mathbf{W}^T\mathbf{x}_i^T - 2\sigma^2_{\delta,B-GLS}\mathbf{x}_i\mathbf{W}e_i\right] \qquad (3\text{-}13b)$$

$$VP_{old}(i) = E\left[\sigma^2_{\delta,B-GLS}\right] + \mathbf{x}_i Var\left[\hat{\beta}_{WLS}\right]\mathbf{x}_i^T - 2\left(E\left[\sigma^2_{\delta,B-GLS}\right]\right)\mathbf{x}_i\mathbf{W}e_i$$

where e_i is the $(n \times 1)$ column vector with one at the ith row and zero otherwise.

Leverage

The leverage measure, $\mathbf{H}^*$, for a GLS regression, as described by Tasker and Stedinger (1989, eq. 23), is calculated as follows:

$$\mathbf{H}^* = \mathbf{X}\left\{\mathbf{X}^T\left(\mathbf{\Lambda_{GLS}}\left(\sigma^2_{\delta,MM-GLS}\right)\right)^{-1}\mathbf{X}\right\}^{-1}\mathbf{X}^T\left(\mathbf{\Lambda_{GLS}}\left(\sigma^2_{\delta,MM-GLS}\right)\right)^{-1} \qquad (3\text{-}14)$$

With the WLS/GLS methodology used in this study, the WLS step selects weights, $\mathbf{W}$, to be used to estimate the coefficients and, thus, determines the leverage that should be associated with each observation. In calculating the leverage, a diagonal covariance matrix is used with the B-WLS model error variance. Thus, by using the framework for leverage provided by Tasker and Stedinger (1989), the leverage for this study is as follows:

$$\mathbf{H}^*_{WLS} = \mathbf{XW} \qquad (3\text{-}15)$$

or

$$\mathbf{H}^*_{WLS} = \mathbf{X}\left\{\mathbf{X}^T\left(\mathbf{\Lambda}_{WLS}\left(\sigma^2_{\delta,B-WLS}\right)\right)^{-1}\mathbf{X}\right\}^{-1}\mathbf{X}^T\left(\mathbf{\Lambda}_{WLS}\left(\sigma^2_{\delta,B-WLS}\right)\right)^{-1}$$

where $\mathbf{\Lambda_{WLS}}\left(\sigma^2_{\delta,B-WLS}\right)$ is an $(n \times n)$ covariance matrix, described in equation 3–6, in which $\sigma^2_{\delta,B-WLS}$ is the mean model error variance estimated by B-WLS.

Influence

The influence measure, $\mathbf{D}^*$, for a GLS analysis, as proposed by Tasker and Stedinger (1989, eqs. 25–26), is a generalized form of the Cook's D and was computed as follows:

$$D_i^* = \frac{1}{k}\frac{\left[\mathbf{H}^*\left(\mathbf{\Lambda_{GLS}}\left(\sigma^2_{\delta,MM-GLS}\right)\right)\right]_{ii}\hat{\varepsilon}_i^2}{\left[\left(\mathbf{I}-\mathbf{H}^*\right)\left(\mathbf{\Lambda_{GLS}}\left(\sigma^2_{\delta,MM-GLS}\right)\right)\right]_{ii}^2} \qquad (3\text{-}16)$$

where k is the number of estimated regression coefficients, $\hat{\varepsilon}_i$ is the residual error for site i, $\mathbf{H}^*$ is an $(n \times n)$ matrix of the GLS leverage, $\mathbf{\Lambda_{GLS}}\left(\sigma^2_{\delta,MM-GLS}\right)$ is an $(n \times n)$ covariance matrix, and $\mathbf{I}$ is an $(n \times n)$ identity matrix. Equation 3–16 can be simplified:

$$D_i^* = \frac{h_{ii}'\hat{\varepsilon}_i^2}{k\left(\lambda_{ii}' - h_{ii}'\right)^2} \qquad (3\text{-}17)$$

where h_{ii}' are the diagonal elements of the following:

$$\mathbf{H}' = \mathbf{H}^* \left(\mathbf{\Lambda}_{GLS} \left(\sigma_{\delta, MM-GLS}^2 \right) \right) = \mathbf{X} \left\{ \mathbf{X}^T \left(\mathbf{\Lambda}_{GLS} \left(\sigma_{\delta, MM-GLS}^2 \right) \right)^{-1} \mathbf{X} \right\}^{-1} \mathbf{X}^T \tag{3-18}$$

and λ_{ii}' is the ith diagonal element of $\mathbf{\Lambda}_{GLS} \left(\sigma_{\delta, MM-GLS}^2 \right)$.

The influence metric adopted by Tasker and Stedinger (1989) needs to be recast for the WLS/GLS methodology used in this study. Here, the regression coefficients are estimated by using WLS, whereas the precision of those coefficients and the precision of the model are calculated by using Bayesian GLS.

As shown in equation 3–16, Cook's **D** contains two terms. The first describes the leverage of a point, which is measured as $Var[\hat{\gamma}_i \,|\, WLS \text{ model}]/Var[\hat{\varepsilon}_i \,|\, WLS \text{ model}]$, and the second is the square of the residual error divided by its variance.

The values of the required variance follow. In this formulation, $\mathbf{\Lambda} = \mathbf{\Lambda}_{GLS} \left(\sigma_{\delta, B-GLS}^2 \right)$, $\mathbf{L} = \mathbf{\Lambda}_{WLS} \left(\sigma_{\delta, B-WLS}^2 \right)$, and $\mathbf{H}_{WLS}^* = $ WLS/GLS Leverage (see eq. 3–15).

This is done to simplify the following equations:

$$\begin{aligned} Var[\hat{\gamma} \,|\, WLS \text{ model}] &= \left(\mathbf{H}_{WLS}^* \right) \mathbf{L} \left(\mathbf{H}_{WLS}^* \right) \\ &= \mathbf{X} \mathbf{W}_{WLS} \mathbf{L} \mathbf{W}_{WLS}^T \mathbf{X}^T \\ &= \mathbf{X} \left(\mathbf{X}^T \mathbf{L}^{-1} \mathbf{X} \right)^{-1} \mathbf{X}^T \mathbf{L}^{-1} \mathbf{L} \mathbf{L}^{-1} \mathbf{X} \left(\mathbf{X}^T \mathbf{L}^{-1} \mathbf{X} \right)^{-1} \mathbf{X}^T \\ &= \mathbf{X} \left(\mathbf{X}^T \mathbf{L}^{-1} \mathbf{X} \right)^{-1} \mathbf{X}^T = \mathbf{X} \mathbf{W} \mathbf{L} = \left(\mathbf{H}_{WLS}^* \right) \mathbf{L} \end{aligned} \tag{3-19}$$

$$\begin{aligned} Var[\hat{\varepsilon} \,|\, WLS \text{ model}] &= \mathbf{E} \left\{ \left(\boldsymbol{\gamma} - \left(\mathbf{H}_{WLS}^* \right) \boldsymbol{\gamma} \right) \left(\boldsymbol{\gamma} - \left(\mathbf{H}_{WLS}^* \right) \boldsymbol{\gamma} \right)^T \right\} \\ &= \mathbf{L} - \left(\mathbf{H}_{WLS}^* \right) \mathbf{L} - \mathbf{L} \left(\mathbf{H}_{WLS}^* \right)^T + \left(\mathbf{H}_{WLS}^* \right) \mathbf{L} \left(\mathbf{H}_{WLS}^* \right)^T \end{aligned} \tag{3-20}$$

$$\begin{aligned} Var[\hat{\varepsilon} \,|\, GLS \text{ model}] &= \mathbf{E} \left\{ \left(\boldsymbol{\gamma} - \left(\mathbf{H}_{WLS}^* \right) \boldsymbol{\gamma} \right) \left(\boldsymbol{\gamma} - \left(\mathbf{H}_{WLS}^* \right) \boldsymbol{\gamma} \right)^T \right\} \\ &= \mathbf{\Lambda} - \left(\mathbf{H}_{WLS}^* \right) \mathbf{\Lambda} - \mathbf{\Lambda} \left(\mathbf{H}_{WLS}^* \right)^T + \left(\mathbf{H}_{WLS}^* \right) \mathbf{\Lambda} \left(\mathbf{H}_{WLS}^* \right)^T \end{aligned} \tag{3-21}$$

$$D_i^{WG} = \left(\frac{1}{k} \right) \left(\frac{Var[\hat{\gamma}_i \,|\, WLS \text{ model}]}{Var[\hat{\varepsilon}_i \,|\, WLS \text{ model}]} \right) \left(\frac{\varepsilon_i^2}{Var[\hat{\varepsilon}_i \,|\, GLS \text{ model}]} \right) \tag{3-22}$$

or

$$D_i^{WG} = \left(\frac{1}{k} \right) \left(\frac{h_{WLS,ii}^*}{1 - h_{WLS,ii}^*} \right) \left(\frac{\varepsilon_i^2}{Var[\hat{\varepsilon}_i \,|\, GLS \text{ model}]} \right)$$

Here, $h_{WLS,ii}^*$ are the diagonal elements of H_{WLS}^*. The influence metric, described by equation 3–22 takes into account the mixed WLS/GLS analysis used to generate the regional-skew model. The predicted regional-skew model is estimated by using WLS, and, thus, the leverage metric reflects the WLS weights that depend on the diagonal covariance matrix. However, GLS describes the actual precision of the model and the precision of the residuals. Thus, the last term in equation 3–15 uses the correct estimate of the variance of the computed residuals as computed by the GLS analysis.

Leverage and Influence for Sites in the Regional-Skew Analysis for Rainfall Floods of Differing Durations

Equations 3–15 and 3–22 provide the leverage and influence values for each site and duration included in the WLS/GLS regression analyses. If $\hat{\beta}$ has dimensionality k and n is the sample size (number of basins in the study), the mean of the leverage values is k/n; thus, values greater than $2k/n$ are generally considered to be large. Influence values greater than $4/n$ are typically considered to be large (Stedinger and Tasker, 1985). By using these relationships, in this study, leverage values greater than 0.12 were considered to be large, and influence values greater than 0.08 were considered to be large.

Figure 3-1 shows influence and leverage statistics for each site for 1-day, 3-day, 7-day, 15-day, and 30-day durations. Leverage values did not change radically from one duration to another because the matrix of basin characteristics and the sample sizes were the same for all durations; however, the model error variances were different, which resulted in some differences in the leverage values. On the other hand, the influence values depended on the residuals computed from the individual skew regressions for each duration and, thus, changed from one duration to another. None of the basins in this study had high leverage values at any duration. Furthermore, no more than three basins showed high influence for any duration. Those basins whose influence did exceed 0.08 did not exceed it by much, so their influences were not large enough to be alarming. No basin had high influence at all durations.

South Fork Eel River near Miranda (site 46) had high influence at shorter durations (1 day and 3 days). With 68 years of record, this site had a very high residual for the 1-day and 3-day durations. The influence, particularly for 1 day, was only marginally high and therefore not significant to the study.

The Tuolumne River at New Don Pedro Dam (site 25) and the Kings River at Pine Flat Dam (site 18) had high influence at longer durations (15 days and 30 days). These sites have a long record length (112 years and 113 years, respectively) which results in larger weights and, thus, relatively high leverage. Since the residuals are large at the longer durations and those sites have long record lengths, it was not surprising that the influences were very high.

Upon inspection of leverage and influence values for this study, the Sacramento River at Shasta Dam (site 1) had very high influence for the 1-day duration but only moderate and low influence at other durations. This warranted a closer examination of the 1-day duration for this site. It was found that a low outlier was not censored, as it should have been, during the frequency analysis. This resulted in an uncharacteristically negative skew in the frequency curve. After this additional value was censored and the regression was re-run, the fit of the regional-skew model improved, and the influence of the 1-day duration at this site was reasonable. Usually, it is not a good practice to change the number of observations censored explicitly to achieve a desired result, but in this case, the diagnostic statistics alerted the researchers to an error in a previous analysis, which after correction, by happenstance, resulted in an improved model.

Overall, site 1 is an example where large leverage values were not expected. The value of the nonlinear function of elevation ranged from zero for basins below 3,000 feet to one for basins above 4,200 feet. Thus, it was impossible for any basin to have an extreme value. Sampling error associated with each skew coefficient also contributed to the leverage. The longer-record sites did not have record lengths much longer than 100 years, and many sites had record lengths about that long. Thus, no sites were unusual. Examining the final leverage and influence statistics indicated there were no problems in the development of the flood data, the basin characteristics file, the at-site skew estimators, or in the statistical analyses.

References

Griffis, V.W., and Stedinger, J. R., 2009, Log-Pearson type 3 distribution and its application in flood frequency analysis, III: sample skew and weighted skew estimators: Journal of Hydrology, v. 14, no. 2, p. 121–130.

Parrett, C., Veilleux, A., Stedinger, J.R., Barth, N.A., Knifong, D.L., and Ferris, J.C., 2011, Regional skew for California, and flood frequency for selected sites in the Sacramento–San Joaquin River Basin, based on data through water year 2006: U.S. Geological Survey Scientific Investigations Report 2010–5260, 94 p.

Reis, D.S., Jr., Stedinger, J.R., and Martins, E.S., 2005, Bayesian generalized least squares regression with application to the log Pearson type III regional skew estimation: Water Resources Research, v. 41, W10419, doi:10.1029/2004WR003445.

Tasker, G.D., and Stedinger, J.R., 1986, Regional skew with weighted LS regression: Journal of Water Resources Planning and Management, American Society of Civil Engineers, v.112, no. 2, p. 225–237.

Tasker, G.D., and Stedinger, J.R., 1989. An operational GLS model for hydrologic regression: Journal of Hydrology, v. 111, no. 1–4, p. 361–375.

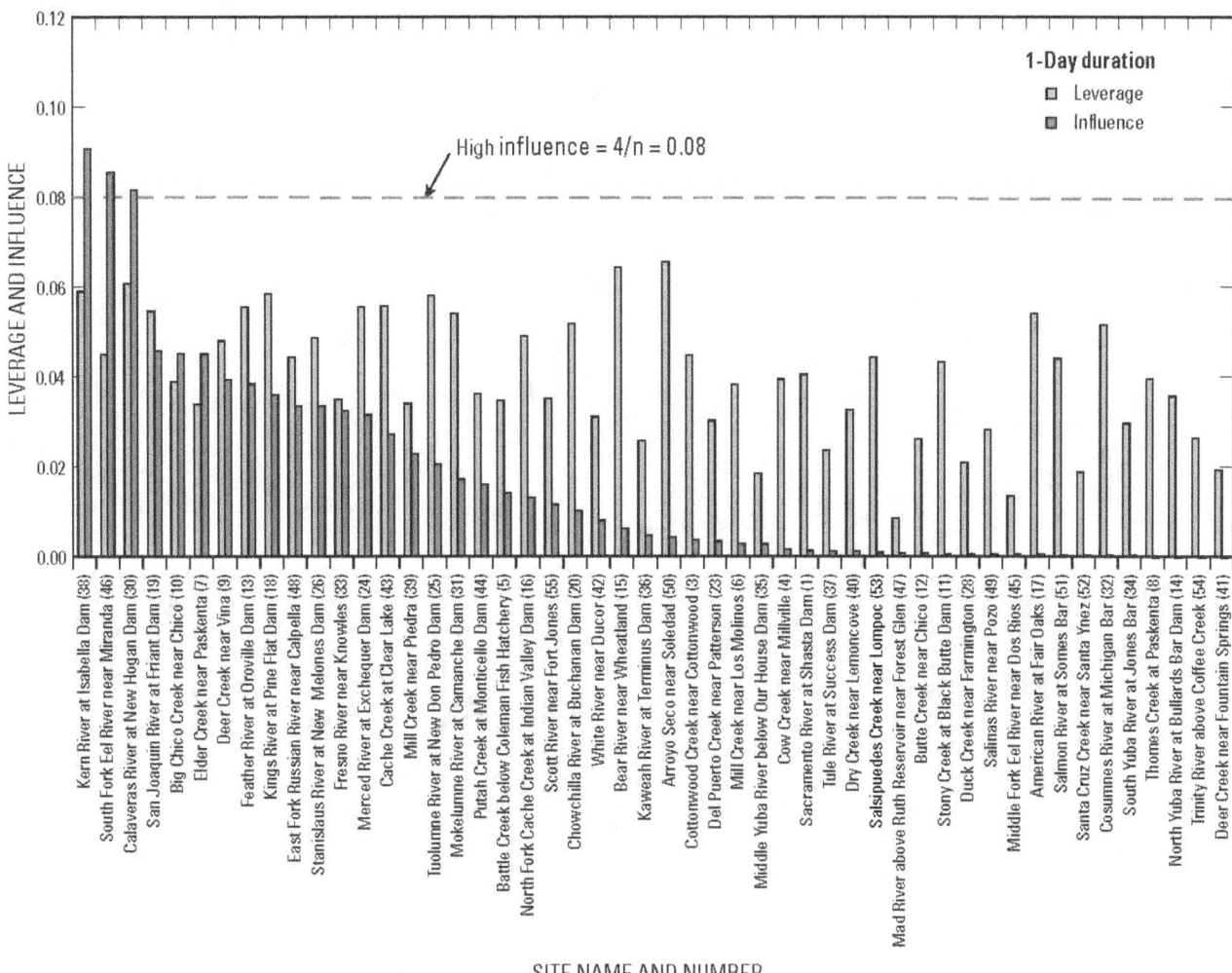

Figure 3–1. Leverage and influence for study sites for (*A*) 1-day, (*B*) 3-day, (*C*) 7-day, (*D*) 15-day, and (*E*) 30-day duration, where *n* represents the number of sites.

B

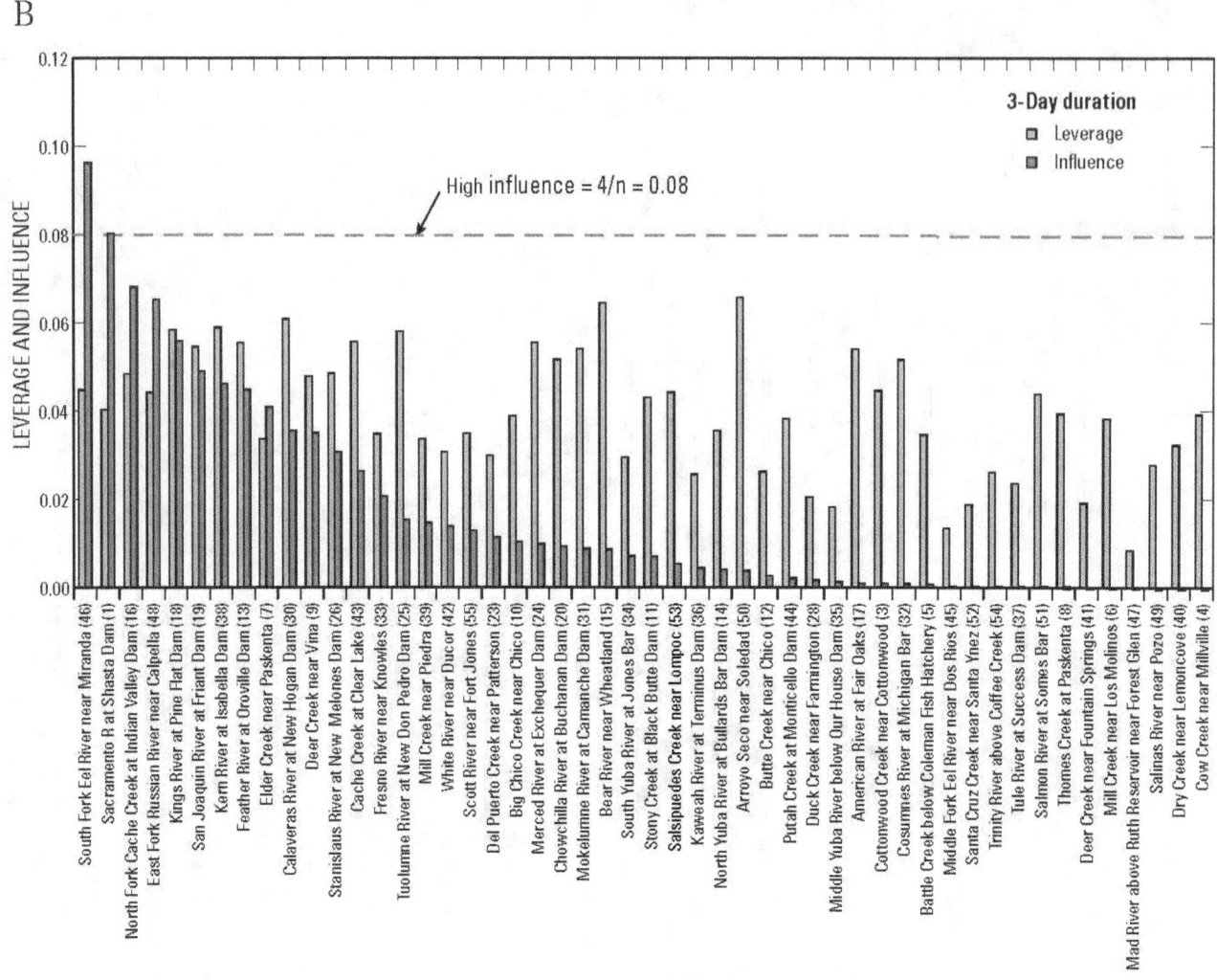

SITE NAME AND NUMBER

Figure 3–1.—Continued

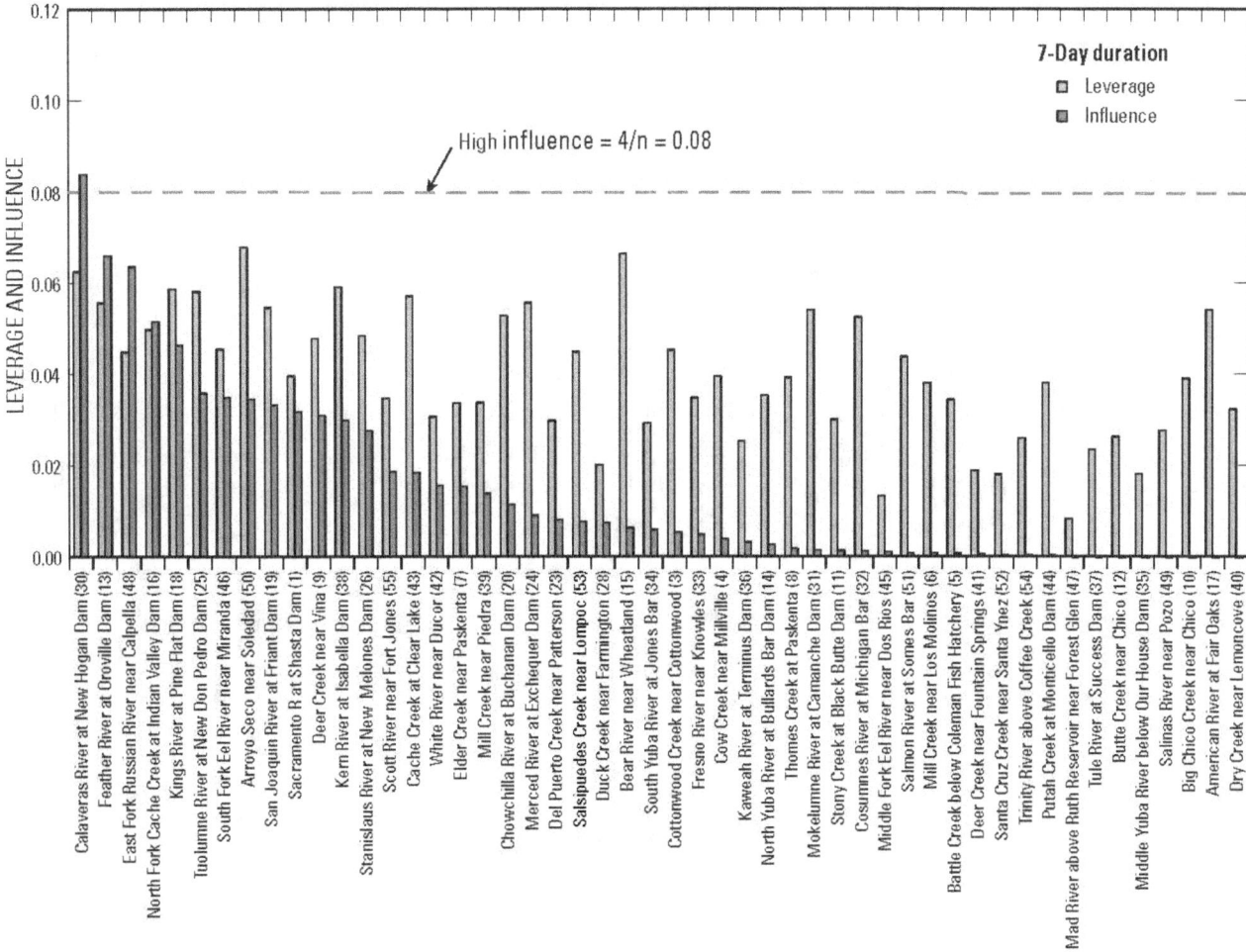

SITE NAME AND NUMBER

Figure 3–1.—Continued

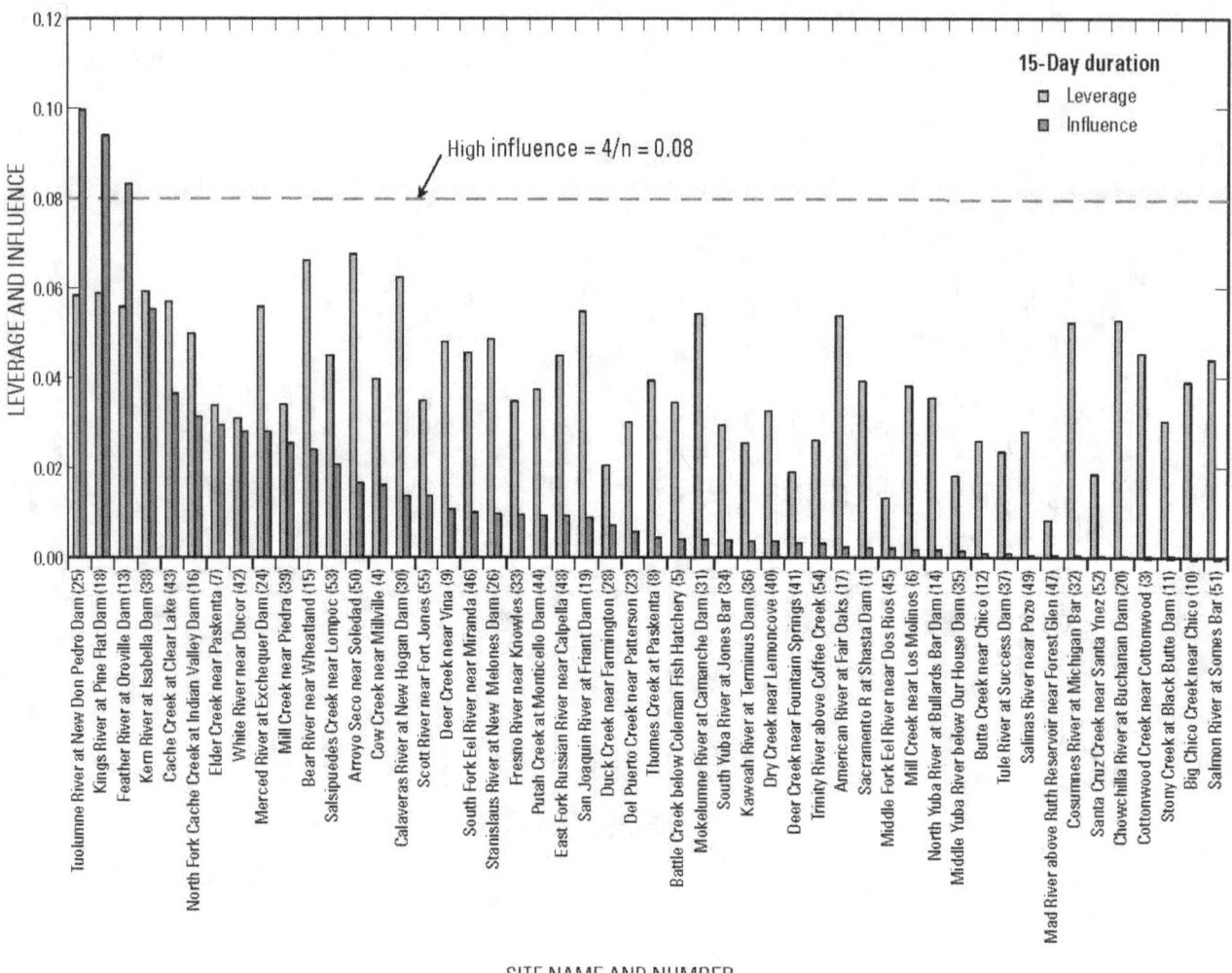

SITE NAME AND NUMBER

Figure 3–1.—Continued

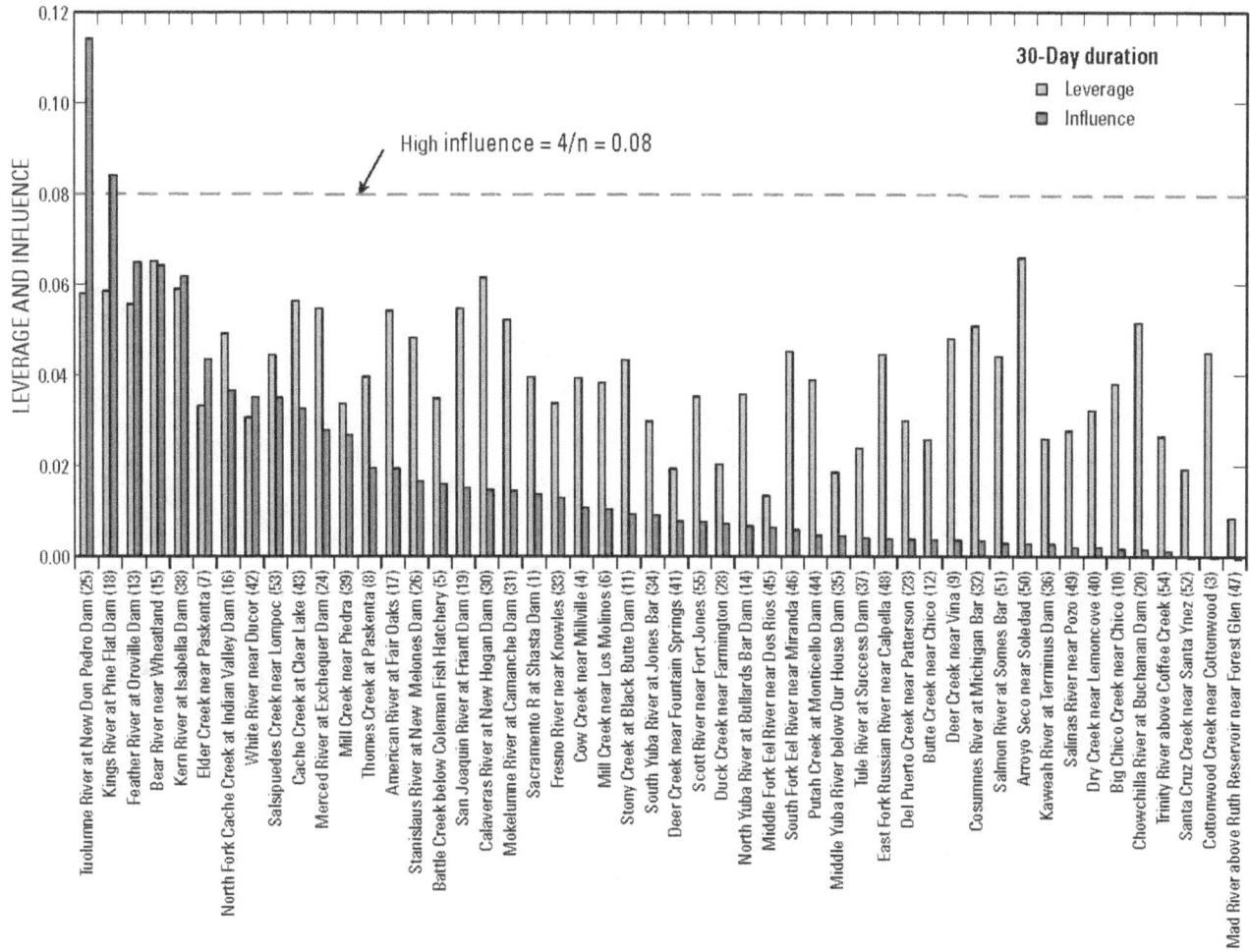

SITE NAME AND NUMBER

Figure 3–1.—Continued

www.ingramcontent.com/pod-product-compliance
Lightning Source LLC
Chambersburg PA
CBHW081607170526
45166CB00009B/2859